The Logica Yearbook
2023

The Logica Yearbook 2023

Edited by

Igor Sedlár

© Individual authors and College Publications 2025
All rights reserved.

ISBN 978-1-84890-032-5

College Publications
Scientific Director: Dov Gabbay
Managing Director: Jane Spurr

www.collegepublications.co.uk

Original cover design by Laraine Welch

All rights reserved. No part of this publication may be reproduced, stored in a retrieval system or transmitted in any form, or by any means, electronic, mechanical, photocopying, recording or otherwise without prior permission, in writing, from the publisher.

Preface

This volume contains peer-reviewed papers based on selected contributions presented at the conference *Logica 2023*, which took place in the Teplá Monastery, Czech Republic, on 18–22 June 2023. Since 1987 the Logica conference series consists of meetings typically taking place in a quiet venue housing all participants together, and fostering fruitful exchanges between logicians of different specializations and generations, including students.

The programme of Logica 2023 comprised more than thirty lectures, including those given by our distinguished invited speakers David Corfield, Wesley Holliday, Gillian Russell, and Sara Uckelman. On behalf of the Logica co-chair Vít Punčochář and the whole Logica organising team, I would like to thank the Institute of Philosophy of the Czech Academy of Sciences for its support of the Logica conference series, and the staff of Teplá Monastery Hotel for their hospitality and friendly assistance. Vít and I are grateful to the members of the programme committee for evaluating the abstracts submitted to the conference, to Anna Kratochvílová for administrative and practical assistance, and to Ondrej Majer for designing the Logica T-shirts and other materials. As the editor of this volume, I am grateful to the reviewers of the papers for their time and valuable advice, and to College Publications and its managing director, Jane Spurr, for our pleasant cooperation during the preparation of this volume. Last but not least, I would like to thank authors of the papers included in this volume for their contributions and collaboration during the editorial process.

Prague, March 2025   Igor Sedlár

Table of Contents

γ-Admissibility in Mares' Varying Domain **QR**.................... 1
  Nicholas Ferenz and Thomas M. Ferguson

A Uniform Approach to Variable-Sharing in Relevant Logics 17
  Thomas M. Ferguson and Jitka Kadlečíková

Reconsidering Identity.. 37
  René Gazzari

Preconditionals 59
  Wesley H. Holliday

Aristotle's Syllogistic Logic as a Theory of Arithmetical Kind 79
  Ladsilav Kvasz

Dialectical Dispositions and Logic 103
  Lionel Shapiro

Generalized Bilateral Harmony 123
  Ryan Simonelli

Logic as Liberation, or, Logic, Feminism, and Being a Feminist
in Logic 145
  Sara L. Uckelman

γ-Admissibility in Mares' Varying Domain QR

NICHOLAS FERENZ[1] AND THOMAS M. FERGUSON[2]

Abstract: In his *General Information in Relevant Logic*, Edwin Mares provided an informational interpretation of situation-based semantics for relevant logic, motivating the interpretation by offering several detailed philosophical examples. Noting that several of these examples presupposed an individual's existing at some situations but not at others, Mares concluded the work by introducing a varying-domain version of the quantified relevant logic **QR**. In this paper, we use the technique of *normalization* to establish γ-admissibility for Mares' varying-domain **QR**.

Keywords: relevant logic, varying-domains, γ-admissibility

1 Introduction

The status of Ackermann's rule γ described by Ackermann (1956)—*i.e.* from $\neg\varphi \vee \psi$ and φ infer ψ—has traditionally been a matter of great importance in the field of mainstream relevant logics. The admissibility of γ, i.e., the closure of the set of theorems of a logic under the rule, has often been of great importance in the philosophical defense of relevant logics. Recent work has been devoted to examining the admissibility of γ in expansions of propositional relevant logics, *e.g.*, modal expansions (*e.g.* (Seki, 2011a) and (Seki, 2011b)) and first-order expansions (*e.g.* (Ferenz & Ferguson, 2023)).

Mares (2009) introduces a varying-domain first-order expansion of the relevant logic **R** two features of which make it interesting from the standpoint of such work. For one, the language includes an existence predicate E leveraged in the definition of the universal quantifier. For two, Mares' model theory for the system makes use of a compatibility relation C in the style of

---

[1] The first author thanks the organizers of Logica, and the Logica attendees and anonymous referees for the comments which undoubtedly improved this paper. This paper was supported by RVO 67985807.

[2] The second author thanks the participants of Logica for invaluable comments and feedback.

(Dunn, 1993). Both devices are uncommon enough to require additional care when considering γ admissibility.

In this paper, we prove γ admissibility for this expansion, using the method of normalization described in (Routley, Plumwood, Meyer, & Brady, 1982) and developed further *e.g.* in (Seki, 2011a). Like the companion paper (Ferenz & Ferguson, 2023), this work improves on Seki's paper by accounting for elements in the Mares-Goldblatt-style model theory like the set of propositions *Prop*. We will first introduce Mares' varying-domain **QR** and then consider how the technique can be adapted in order to show the admissibility of γ.

2 Mares' varying-domain QR

In (Mares, 2009), Edwin Mares provides an in-depth analysis of the informational aspects of relevant logic, focusing on providing an informational interpretation of the semantics for quantified relevant logic presented in (Mares & Goldblatt, 2006).

In the concluding sections of the book, Mares notes that certain informal illustrations of this informational interpretation make assumptions that contradict the model theory of **QR**, namely, that there may be two distinct situation one of which assumes the existence of some individual and the other one of which does not.

> In the [discussion], we appeal to the idea that certain entities can be present in a situation and others not present. In order to make sense of this distinction in our semantics, we need to have the domains of individuals vary from situation to situation.(Mares, 2009, p. 435)

This poses a dilemma: The utility of Mares' philosophical interpretation relies on the representational adequacy of the details of the model theory for relevant logics but the portrait of quantification described in (Mares & Goldblatt, 2006) assumes the existence of a single domain of individuals shared by all states.

Mare's solution is to define a first-order expansion of **R** including an existence predicate through which the universal quantifier gains existential import. We will now consider this system—which we describe as *Mares' varying-domain* **QR**—by providing proof and model theories in the following sections.

2.1 Axiomatization of Mares' system

Mares' varying-domain **QR** is given both natural deduction and axiomatic presentations as an expansion of **R**. The following axiomatic presentation is found in (Mares, 2009):

1. $\varphi \to \varphi$
2. $(\varphi \to \psi) \to ((\psi \to \xi) \to (\varphi \to \xi))$
3. $\varphi \to ((\varphi \to \psi) \to \psi)$
4. $(\varphi \to (\varphi \to \psi)) \to (\varphi \to \psi)$
5. $(\varphi_0 \wedge \varphi_1) \to \varphi_i \ (i \in \{0,1\})$
6. $\varphi_i \to (\varphi_0 \vee \varphi_1) \ (i \in \{0,1\})$
7. $((\varphi \to \psi) \wedge (\varphi \to \xi)) \to (\varphi \to (\psi \wedge \xi))$
8. $((\varphi \vee \psi) \to \xi) \leftrightarrow ((\varphi \to \xi) \wedge (\psi \to \xi))$
9. $(\varphi \wedge (\psi \vee \xi)) \to ((\varphi \wedge \psi) \vee (\varphi \wedge \xi))$
10. $(\varphi \to \neg\psi) \to (\psi \to \neg\varphi)$
11. $\neg\neg\varphi \to \varphi$
12. $\forall x \varphi \to (Et \to \varphi[t/x])$ where x is free for t

R1. $\varphi \to \psi, \varphi \Rightarrow \psi$
R2. $\varphi, \psi \Rightarrow \varphi \wedge \psi$
R3. $\varphi^x \to (Ex \to \psi) \Rightarrow \varphi^x \to \forall x \psi$ where x not free in φ

Derivability for Mares' system can be provided in the standard way from these axioms.

2.2 Model theory for Mares' system

We now proceed to examine the model theory for Mares' varying-domain **QR**, which largely resembles the work in (Mares & Goldblatt, 2006). We first introduce the basic component of the model-theoretic construction:

Definition 1 *A model structure is a tuple* $\mathcal{S} = \langle S, N, R, C, I, D, \cdot^*, Prop \rangle$ *where:*

- *S is a nonempty set of situations*

- *$N \subseteq S$ is a non-empty set of* logical *situations*

- *R is a ternary relation on S*

- *C is a binary compatibility relation on S*

- *I is a non-empty set of individuals*

- *D is a unary function from S to $\wp(I)$*

- $\cdot^*$ *is a unary function from S to S*
- *$Prop$ is a subset of $\wp(S)$*

Define the binary relation $s \sqsubseteq t$ so that $s \sqsubseteq t$ iff there exists an $x \in N$ such that $Rxst$ and call an $X \in \wp(W)$ an upset *if X is upwards closed under $\sqsubseteq$. Then we insist on the following provisions:*

C1 $\sqsubseteq$ *is transitive and reflexive*

C2 $Rsss$

C3 *If $Rstu$ then $Rtsu$*[3]

C4 *If $Rstu$ and $s' \sqsubseteq s$ then $Rs'tu$*

C5 *If for some $a \in S$, $Rsta$ & $Rauv$ then for some $b \in S$ s.t. $Rsub$ & $Rbtv$*

C6 *For all s: Css^* and for all t such that Cst, $t \sqsubseteq s^*$*

C7 *If $Rstu$ then Rsu^*t^**

C8 $s^{**} = s$

C9 $N \in Prop$

C10 *For all $\pi \in Prop$, π is a $\sqsubseteq$-upset*

C11 *For all s, the set $\{t \mid Cst\}$ is a $\sqsubseteq$-upset*

C12 *$Prop$ is closed under $\neg$, $\cap$, $\cup$, and $\rightarrow$.*

where operations $\neg$ and $\rightarrow$ on elements of $Prop$ are defined as $\neg X = \{s \in S \mid s^* \notin X\}$ and $X \rightarrow Y = \{s \in S \mid \text{for all } t, u \in S \text{ if } Rstu \text{ and } t \in X \text{ then } u \in Y\}$.

Mares incrementally introduces the notion of model that is built upon a model structure. Citing the work of Goldblatt and Hodkinson (2009) as inspiration, Mares first introduces the notion of a *premodel*, with the definition of model proper as a premodel of a particular kind. The former is defined as follows:

---

[3] N.b. that in (Mares, 2009), this condition is offered as $Rstu$ implies $Rsut$. We gather that this is a typo. Not only is this condition overly strong, but the justification for the condition is the validation of the axiom $\varphi \rightarrow ((\varphi \rightarrow \psi) \rightarrow \psi)$, which corresponds to the frame condition as we have presented it.

Definition 2 *A premodel is a pair* $\mathcal{M} = \langle \mathcal{S}, |\cdot|^\mathcal{M} \rangle$ *with $\mathcal{S}$ a model structure and $|\cdot|^\mathcal{M}$ is defined so that:*

- *for each constant c in the signature, $|c|^\mathcal{M} \in I$*
- *for an n-ary relation P in the signature, $|P|^\mathcal{M} : I^n \to Prop$*

The notion of model will be introduced as a premodel that evaluates all formulae in an appropriate way, *i.e.*, as *propositions* (or members of $Prop$). To pin the requirements down, we must introduce several definitions.

An *assignment* f is a function from the set of variables $\{x_n \mid n \in \mathbb{N}\}$ to I. We let the notation $f[i/n]$

$$f[i/n](x_m) = \begin{cases} i \text{ if } m = n \\ f(x_m) \text{ if } m \neq n \end{cases}$$

Assignments serve as devices to interpret syntax, beginning with terms. For a term t,

$$t^\mathcal{M} f = \begin{cases} |t|^\mathcal{M} \text{ if } t \text{ is a constant} \\ f(x_n) \text{ if } t \text{ is the varying-}x_n \end{cases}$$

This lifts to provide a definition of evaluations for the entire language:

- $|E(t)|^\mathcal{M} f = \{s \in S \mid t^\mathcal{M} f \in D(s)\}$
- $|P(t_0, ..., t_{n-1})|^\mathcal{M} f = |P|^\mathcal{M}(t_0^\mathcal{M} f, ..., t_{n-1}^\mathcal{M} f)$
- $|\varphi \wedge \psi|^\mathcal{M} f = |\varphi|^\mathcal{M} f \cap |\psi|^\mathcal{M} f$
- $|\varphi \vee \psi|^\mathcal{M} f = |\varphi|^\mathcal{M} f \cup |\psi|^\mathcal{M} f$
- $|\neg \varphi|^\mathcal{M} f = \{s \in S \mid \text{ for all } t \in S, \text{ if } Cst \text{ then } t \notin |\varphi|^\mathcal{M} f\}$
- $|\varphi \to \psi|^\mathcal{M} f = \{s \in S \mid \text{ for all } t, u \in S \text{ if } Rstu \;\&\; t \in |\varphi|^\mathcal{M} f, \text{ then } u \in |\psi|^\mathcal{M} f\}$
- $|\forall x_n \varphi|^\mathcal{M} f = \sqcap_{i \in I}(|E(x_n)| \to \varphi|^\mathcal{M} f[i/n])$

where $\sqcap_{i \in I} X = \cup \{\pi \in Prop \mid \pi \subseteq \cap X\}$.

Having spelled out what it means for a formula to be interpreted, we can now precisely introduce the definition of model:

Definition 3 *A model $\mathcal{M}$ is a premodel in the sense of Definition 2 such that for all assignments f and formulae φ, $|\varphi|^{\mathcal{M}} f \in Prop$.*

Before proceeding, let us make a quick observation. Mares' use of the compatibility relation is borrowed from Dunn (1993) but deviates from the more familiar treatment of negation using the Routley star. In order to simplify matters, we will at some points appeal to the following fact that aligns the compatibility relation C with the Routley star $\cdot^*$:

Lemma 1 *In a premodel:*

$$\{s \mid \text{for all } t, Cst \Rightarrow t \notin |\varphi|^{\mathcal{M}}\} = \{s \mid s^* \notin |\varphi|^{\mathcal{M}}\}.$$

Proof. For right-to-left, pick an s and suppose that $s^* \notin |\varphi|^{\mathcal{M}}$. Then as s^* is a $\sqsubseteq$-greatest element of $\{t \mid Cst\}$, for any t such that Cst, $t \sqsubseteq s^*$, whence if φ fails at s^*, φ fails at t. As t was arbitrary, for all t, $Cst \Rightarrow t \notin |\varphi|^{\mathcal{M}}$. For left-to-right, suppose that for all t, if Cst then $t \notin |\varphi|^{\mathcal{M}}$. Then, insofar as Css^* holds, *a fortiori*, $s^* \notin |\varphi|^{\mathcal{M}}$. □

Definition 4 *A formula φ is valid if for any model, for all assignments f, $N \subseteq |\varphi|^{\mathcal{M}} f$*

Now, we have the following courtesy of (Mares, 2009):

Theorem 2 (Mares) *Mares' varying-domain* **QR** *is sound and complete with respect to the foregoing semantics.*

3 γ-admissibility in Mares' system

To show γ-admissibility, we use the method of normal models introduced in (Routley & Meyer, 1972). This requires that we first introduce the central property of *normality*:

Definition 5 *A model structure (respectively, premodel or model) is normal if there exists an $s \in N$ such that $s = s^*$.*

Remarkably, for any varying-domain **QR** model, we can pick an $o \in N$ and manufacture a *normalized* model by adding a new world 0 relative to o such that $0 = 0^*$.

We will introduce the method of normalization iteratively, starting with introducing the normalization of a model structure, and continuing to premodels and models, respectively. To begin, we define the normalization of a model structure:

Definition 6 *Let S be a model structure with a situation $o \in N$ and let 0 be a situation not in S. Then S the normalization of S at 0 for o is built so that the model structure $S' = \langle S', N', R', C', I', D', \cdot^{*'}, Prop' \rangle$ is defined so that:*

- $S' = S \cup \{0\}$

- $N' = N \cup \{0\}$

- *R' is defined s.t. for all $s, t, u \in S$, $Rstu$ iff $R'stu$ & for all $s, t \in S$*

  - $R'00s$ iff $R'oos$ 　　○ $R'0s0$ iff $R'oso^*$ 　　○ $R'000$
  - $R's00$ iff $R'soo^*$ 　　○ $R'0st$ iff $R'ost$
  - $R's0t$ iff $R'sot$ 　　　○ $R'st0$ iff $R'sto^*$

- *C' is for $s, t \in S$: $C'st$ iff Cst. Otherwise, $C'0s$ iff $C'os$ and $C's0$ iff $C'so$*

- $I' = I$

- *for all $s \in S$ $D'(s) = D(s)$ while $D'(0) = D(o)$*

- *for all $s \in S$, $s^{*'} = s^*$ while $0^{*'} = 0$*

- *$Prop'$ includes all $\pi \in Prop$ when $o^* \notin \pi$, and for all $\pi \in Prop$ such that $o \in \pi$, $\pi \cup \{0\} \in Prop'$*

We will use the nomenclature $\sqsubseteq'$ to denote the ordering relativized to R'.

It, of course, is insufficient to merely define this structure; we must demonstrate that it exhibits the correct properties to qualify it as a model structure. We start by proving some observations, including the following lemma that follows from the same line of argument as exhibited in (Routley et al., 1982):

Lemma 3 *For a model structure S' a normalization of S at 0 for o:*

- *R' is well-defined*

- *for all $s, t \in S$:*

  - $s \sqsubseteq' t$ iff $s \sqsubseteq t$
  - $0 \sqsubseteq' t$ iff $o \sqsubseteq t$

– $s \sqsubseteq' 0$ iff $s \sqsubseteq o^*$

Corollary 1 *For $\mathcal{S}'$ a normalization of $\mathcal{S}$ at 0 for o, $o^* \sqsubseteq' 0 \sqsubseteq' o$*

Corollary 2 *For $\mathcal{S}'$ a normalization of $\mathcal{S}$ at 0 for o, it holds that $C'00$.*

We now proceed to prove that the provisions imposed on model structures by Definition 1 are satisfied by a normalization.

3.1 Adequacy of normalizations

We now show that the technique of normalization yields structures of an appropriate kind. We lead off this section by offering a sequence of lemmas that iteratively show that the conditions in Definition 1 are each satisfied before then treating Definitions 2 and 3.

In the following lemmas, let $\mathcal{S}'$ be the normalization of a model structure $\mathcal{S}$ at 0 for o:

Lemma 4 (Condition **C1**) *The relation $\sqsubseteq'$ is transitive and reflexive.*

Proof. Reflexivity is simple and follows from Lemma 3. For transitivity, Lemma 3 ensures that $s \sqsubseteq' t$ and $t \sqsubseteq' u$ entails $s \sqsubseteq' u$ holds whenever $s, t, u \in S$. The cases in which $t \in S$ are easily covered by appeal to Lemma 3, so consider the case in which $t = 0$. Then $s \sqsubseteq' 0$ and $0 \sqsubseteq' u$ holds iff $s \sqsubseteq' o^*$ and $o \sqsubseteq' u$. □

Lemma 5 (Condition **C2**) *The relation $R'sss$ holds for all $s \in S'$.*

Proof. That $R'sss$ follows from the two cases: If $s \in S$, then this is inherited from $Rsss$. If $s \notin S$ then $s = 0$ but $R'000$ is imposed during construction of the normalization. □

Lemma 6 (Condition **C3**) *If $R'stu$ then $R'tsu$ for all $s, t, u \in S'$.*

Proof. Let $R'stu$. Then let $\bar{s}$ be defined so that $\bar{s} = o$ if $s = 0$ and $\bar{s} = s$ if $s \in S$ and define $\bar{t}$ similarly. Then let $\bar{u} = o^*$ if $u = 0$ and $\bar{u} = u$ otherwise. Then by construction, Lemma 3, and the property's holding in the original model we have: $R'stu$ iff $R\bar{s}\bar{t}\bar{u}$ iff $R\bar{t}\bar{s}\bar{u}$ iff $R'tsu$. □

Lemma 7 (Condition **C4**) *For all $s, s', t, u \in S'$, if $R'stu$ and $s' \sqsubseteq' s$ then $R's'tu$.*

8

Proof. Suppose that $R'stu$. We reduce the number of cases to consider by noting that by construction, each of s, t, u is associated with elements $\bar{s}, \bar{t}, \bar{u}$ in S such that $R'stu$ iff $R\bar{s}\bar{t}\bar{u}$. (E.g., if $t = 0$, let $\bar{t} = o$ and let $\bar{t} = t$ otherwise.) Now, either $s = 0$ or $s \in S$. We consider each case.

- Let $s = 0$. If $s' = 0$, then $s' = s$ and $R's'tu$ follows. So let $s' \in S$. Then we know that $R\bar{s}\bar{t}\bar{u}$ (where $\bar{s} = o$) and that $s' \sqsubseteq \bar{s}$. In the original model, then, $Rs'\bar{t}\bar{u}$, whence by construction $R's'tu$.

- If $s \in S$, then $\bar{s} = s$. So if $s' = 0$, we know that both $Rs\bar{t}\bar{u}$ and $o \sqsubseteq s$, whence in the original model $Ro\bar{t}\bar{u}$ and by construction $R'0tu$, i.e., $R's'tu$. If $s' \in S$, then $s' \sqsubseteq s$ by Lemma 3, and in the original model $Rs'\bar{t}\bar{u}$ holds, whence $R's'tu$ follows. □

Lemma 8 (Condition **C5** (the Pasch Postulate)) *If there is an $a \in S'$ such that $R'sta$ and $R'auv$ then there is a $b \in S'$ such that $R'sub$ and $R'btv$.*

Proof. Define the following for an $s \in S'$:

$$\bar{s} = \begin{cases} o \text{ if } s = 0 \\ s \text{ otherwise} \end{cases} \quad \text{and} \quad \bar{\bar{s}} = \begin{cases} o^* \text{ if } s = 0 \\ s \text{ otherwise} \end{cases}$$

Then suppose that there is an a such that $R'sta$ and $R'auv$. Then by Lemma 3, $R\bar{s}\bar{t}\bar{a}$ and $R\bar{a}\bar{u}\bar{v}$. As $\bar{\bar{a}} \sqsubseteq \bar{a}$ (either trivially, or as $o^* \sqsubseteq o$), by Lemma 7, $R\bar{\bar{a}}\bar{u}\bar{v}$. Consequently, by the Pasch postulate's holding in the original model, there is a $b \in S$ such that $R\bar{s}\bar{u}b$ and $Rb\bar{\bar{t}}\bar{\bar{v}}$. Since $b \neq 0$, we can then infer that $R'sub$ and $R'btv$, as necessary. □

Lemma 9 (Condition **C6**) *For all $s, t \in S'$, $C'ss^{*'}$ and if $C'st$ then $t \sqsubseteq' s^{*'}$.*

Proof. That $C'ss^{*'}$ is immediate by construction. We prove this by examining cases:

- If $s, t \in S$, this follows from Lemma 3 and construction of C'.

- If $s = 0$ and $t \in S$, then we know that in the original model Cot, whence by maximality of o^*, $t \sqsubseteq o^*$. By Lemma 3, $t \sqsubseteq' 0 = 0^{*'}$, whence $t \sqsubseteq' s^{*'}$.

- If $s \in S$ while $t = 0$, then that $C's0$ means that Cso, whence $o \sqsubseteq s^*$. By Lemma 3, $0 \sqsubseteq' s^{*'}$, i.e., $t \sqsubseteq' s^{*'}$.

- If $s, t = 0$ then as $0 = 0^{*'}$, it follows by reflexivity that $t \sqsubseteq' s^{*'}$. □

Lemma 10 (Condition **C7**) *For $s, t, u \in S'$, if $R'stu$ then $R'su^{*'}t^{*'}$.*

Proof. Suppose that $R'stu$. Then we consider the following cases:

- $s, t, u \in S$. Then $R'stu$ iff $Rstu$ iff Rsu^*t^* iff $R'su^{*'}t^{*'}$.

- $s, t \in S$ and $u = 0$. Then $R'st0$ iff $Rsto^*$ iff $Rso^{**}t^*$ iff $Rsot^*$ iff $R's0t^*$. But $0 = 0^{*'}$, whence $R's0^{*'}t^{*'}$, i.e., $R'su^{*'}t^{*'}$.

- $s, u \in S$ and $t = 0$. By analogous reasoning to the case in which $s, t \in S$ and $u = 0$.

- $t, u \in S$ and $s = 0$. Then $R'0tu$ iff $Rotu$ iff Rou^*t^* iff $R'0u^{*'}t^{*'}$, i.e., $R'su^{*'}t^{*'}$.

- $s, t = 0$ and $u \in S$. This follows from similar considerations to $s, u \in S$ and $t = 0$, exchanging s for o in the original model structure.

- $s, u = 0$ and $t \in S$. Follows *mutatis mutandis* from the same argument as the case in which $s, t \in S$ and $u = 0$.

- $t, u = 0$ and $s \in S$. Since $0 = 0^{*'}$, that $R's00$ trivially ensures that $R's0^{*'}0^{*'}$.

- $s, t, u = 0$. Because $0 = 0^{*'}$ $R'000$ is to say that $R'00^{*'}0^{*'}$. □

Lemma 11 (Condition **C8**) *The function $\cdot^{*'}$ is involutive.*

Proof. That $s^{*'*'} = s$ is inherited when $s \in S$; when $s = 0$, this holds from definition of the normalization. □

Lemma 12 (Condition **C9**) $N' \in Prop'$.

Proof. By construction. □

Lemma 13 (Condition **C10**) *Each $\pi \in Prop'$ is $\sqsubseteq'$-closed.*

Proof. Pick an $s \in \pi$ and a t such that $s \sqsubseteq' t$. We show that $t \in \pi$ by considering four cases:

- If $s, t \in S$, as $\pi{\restriction}S \in Prop$ (in the original model structure) and Lemma 3 ensures that $s \sqsubseteq t$, the properties of $\mathcal{S}$ ensure that $t \in \pi$.

- If $s \in S$ but $t \notin S$ then $t = 0$. Thus by Lemma 3, $s \sqsubseteq o^*$, whence since $o^* \sqsubseteq o$, $s \sqsubseteq o$. As $\pi{\restriction}S \in Prop$, $o \in \pi{\restriction}S$ and by construction $0 \in \pi$.

γ-Admissibility in Varying Domain **QR**

- If $s \notin S$ but $t \in S$ then $s = 0$. Thus by construction, $o \in \pi{\restriction}S$ and by Lemma 3, $o \sqsubseteq t$, whence by closure under $\sqsubseteq$ in $\mathcal{S}$, $t \in \pi{\restriction}S$. Consequently, $t \in \pi$.

- If $s \notin S$ and $t \notin S$, then $s = 0$ and $t = 0$, whence $s = t$. Then as $s \in \pi$, that $t \in \pi$ follows trivially. $\square$

Lemma 14 (Condition **C11**)   *For every $s \in S'$, $\{t \mid C'st\}$ is $\sqsubseteq'$-closed.*

Proof. Suppose that $C'st$ and $t \sqsubseteq' u$. We must show that $C'su$. We proceed by examining the following cases:

- $s, t, u \in S$: By Lemma 3 and construction.

- $s, t \in S$ and $u = 0$. Then by Lemma 3, Cst and $t \sqsubseteq o^*$, whence Cso^*. But by Lemma 14 and Corollary 1, $o^* \sqsubseteq o$, whence by closure and Cso. Consequently, $C's0$.

- $s, u \in S$ and $t = 0$. Then by construction, Cso and by Lemma 3, $o \sqsubseteq u$. Thus in the original model structure, Csu, whence $C'su$.

- $t, u \in S$ and $s = 0$. Then in the original model structure, Cot and $t \sqsubseteq u$, whence Cou and $C'ou$, i.e., $C'su$.

- $s \in S$ and $t, u = 0$. As $t = u$, $C'st$ trivially entails $C'su$.

- $t \in S$, and $s, u = 0$. Follows from Corollary 2.

- $u \in S$ and $s, t = 0$. Then from $C'00$ we infer Coo and from $0 \sqsubseteq' u$, infer that $o \sqsubseteq u$, whence Cou in the original model structure. Consequently, by construction $C'0u$, i.e., $C'su$.

- $s, t, u = 0$. Follows from Corollary 2. $\square$

Lemma 15 (Condition **C12**)   *$Prop'$ is closed under $\neg$, $\cap$, $\cup$, and $\to$.*

Proof. Assume that this holds for the original model structure and pick $\pi, \tau \in Prop'$.

- For closure under $\neg$, if $\pi \in Prop'$, then $\pi{\restriction}S \in Prop$ by construction and $\neg \pi{\restriction}S \in Prop$. Either $0 = 0^* \in \pi$ or not. If $0 \in \pi$, then $o \in \pi{\restriction}S$, and thus $o^* \notin \neg \pi{\restriction}S$. If $o \in \neg\pi{\restriction}S$, then $\neg \pi = \neg\pi{\restriction}S \cup \{0\} \in Prop'$; otherwise $\neg \pi = \neg\pi{\restriction}S \in Prop'$. If $0 \notin \pi$, then $o \notin \pi{\restriction}S$. Then $o^* \in \neg\pi{\restriction}S$. So by ordering $o \in \neg\pi{\restriction}S$, and thus $\neg\pi = \neg\pi{\restriction}S \cup \{0\} \in Prop'$.

- For closure under $\cap$, if $\pi, \tau \in Prop'$ then $\pi{\restriction}S \in Prop$ and $\tau{\restriction}S \in Prop$, whence $(\pi \cap \tau){\restriction}S \in Prop$. If $0 \notin \pi \cap \tau$ then $(\pi \cap \tau){\restriction}S = \pi \cap \tau \in Prop'$. If $0 \in \pi \cap \tau$ then $o \in (\pi \cap \tau){\restriction}S$, whence $(\pi \cap \tau){\restriction}S \cup \{0\} = \pi \cap \tau \in Prop'$ by construction. Closure under $\cup$ is by analogous argument.

- For closure under $\to$, fix a $\pi, \tau \in Prop'$. We know that $(\pi{\restriction}S) \to (\tau{\restriction}S) \in Prop$. Suppose first that $0 \in \pi$ and $0 \notin \tau$. By definition $\pi \to \tau = \{u \in S' : \forall v, w \in S'((Ruvw\ \&\ v \in \pi) \to w \in \tau)\}$. Thus, given $Rs0t$ iff $Rsot$, we have $s \in \pi \to \tau$ iff $s \in \pi{\restriction}S \to \tau$, and so $\pi \to \tau \in Prop'$. The remaining cases use $Rst0$ iff $Rsto$ and $Rs00$ iff $Rsoo$. $\square$

Between Lemmas 4–15, we now can establish that the normalization of a model structure is in fact a model structure itself, *i.e.*, that the results of the normalization procedure continue to satisfy Definition 1.

Lemma 16 *Let S' be a normalization of a model structure at 0 for o. Then S' is a model structure in the sense of Definition 1.*

Proof. From construction and Lemmas 4–14. $\square$

Now, let us define the normalization of a premodel.

Definition 7 *For a premodel $\mathcal{M} = \langle \mathcal{S}, |\cdot|^{\mathcal{M}} \rangle$, the premodel $\mathcal{M}' = \langle \mathcal{S}', |\cdot|^{\mathcal{M}'} \rangle$ (the normalization of $\mathcal{M}$ at 0 for o) is defined so that:*

- $|c|^{\mathcal{M}'} = |c|^{\mathcal{M}}$

- $|P|^{\mathcal{M}'}(i^n) = \begin{cases} |P|^{\mathcal{M}}(i^n) & \text{if } o \notin |P|^{\mathcal{M}}(i^n) \\ |P|^{\mathcal{M}}(i^n) \cup \{0\} & \text{if } o \in |P|^{\mathcal{M}}(i^n) \end{cases}$

That the normalization of a premodel at 0 for o is a premodel in the sense of Definition 2 is immediate, given interpretations of constants and predicates. Hence:

Lemma 17 *The normalization $\mathcal{M}'$ of a premodel $\mathcal{M}$ is itself a premodel.*

What is necessary now is to show that the normalization of every *model* is itself a model. This requires an intermediate lemma:

Lemma 18 *Let $\mathcal{M}'$ be the normalization of a premodel $\mathcal{M}$ at 0 for o. Then for all $s \in S$, assignments f, and sentences φ:*

$$s \in |\varphi|^{\mathcal{M}'} f \text{ iff } s \in |\varphi|^{\mathcal{M}} f$$

Proof. We prove this by induction on complexity of φ. As a basis step, if φ is $E(t)$, then $|E(t)|^{\mathcal{M}'}f\restriction S = |E(t)|^{\mathcal{M}}f$ by construction of the normalized model structure $\mathcal{S}'$. Likewise, if φ is $P(t_0,...,t_{n-1})$, $|P(t_0,...,t_{n-1})|^{\mathcal{M}'}f\restriction S = |P(t_0,...,t_{n-1})|^{\mathcal{M}}f$ by construction of the normalized premodel. For the remainder, assume that this holds for all subformulae of φ and suppose that $s \in S$.

- Let $\varphi = \neg\psi$. By Lemmas 1 and 17, $s \in |\neg\psi|^{\mathcal{M}'}f$ iff $s^{*'} \notin |\psi|^{\mathcal{M}'}f$. But $s^{*'} \in S$, so by induction hypothesis, $s^* \notin |\psi|^{\mathcal{M}}f$, which holds iff $s \in |\neg\psi|^{\mathcal{M}}f$.

- Let $\varphi = \psi \wedge \xi$. Then $s \in |\psi \wedge \xi|^{\mathcal{M}'}f$ iff $s \in |\psi|^{\mathcal{M}'}f$ and $s \in |\xi|^{\mathcal{M}'}f$. By induction hypothesis, this is equivalent to $s \in |\psi|^{\mathcal{M}}f$ and $s \in |\xi|^{\mathcal{M}}f$, i.e., that $s \in |\psi \wedge \xi|^{\mathcal{M}}f$. The case in which $\varphi = \psi \vee \xi$ follows from analogous reasoning.

- Let $\varphi = \psi \to \xi$. We prove the left-to-right direction by contraposition, supposing that $s \notin |\psi \to \xi|^{\mathcal{M}'}f$. Then for some $t, u \in S'$ for which $R'stu$ while $t \in |\psi|^{\mathcal{M}'}f$ and $u \notin |\xi|^{\mathcal{M}'}f$. Then, borrowing our notation from the proof of Lemma 8, $R s \bar{t} \bar{u}$, $\bar{t} \in |\psi|^{\mathcal{M}}f$ and $\bar{u} \notin |\xi|^{\mathcal{M}}f$, i.e., $s \notin |\psi \to \xi|^{\mathcal{M}}f$. Hence, if $s \in |\psi \to \xi|^{\mathcal{M}}f$ then $s \in |\psi \to \xi|^{\mathcal{M}'}f$. For the right-to-left direction, suppose that $s \in |\psi \to \xi|^{\mathcal{M}'}f$. If for all $t, u \in S'$ such that $R'stu$, if $t \in |\psi|^{\mathcal{M}'}f$ then $u \in |\xi|^{\mathcal{M}'}$, it follows from Lemma 3 and induction hypothesis that the same can be said for all $t, u \in S$. Hence, $s \in |\psi \to \xi|^{\mathcal{M}}f$.

- Let $\varphi = \forall x_n \psi$. Then $s \in |\forall x_n \psi|^{\mathcal{M}'}f$ iff $s \in \sqcap_{i \in I'}|E(x_n) \to \psi|^{\mathcal{M}'}f[i/n]$, which holds iff $s \in \cup\{\pi \in Prop' \mid \pi \subseteq \sqcap_{i \in I'}|E(x_n) \to \psi|^{\mathcal{M}'}f[i/n]\}$. This holds iff there exists a $\pi \in Prop'$ such that $\pi \subseteq \sqcap_{i \in I'}|E(x_n) \to \psi|^{\mathcal{M}'}f[i/n]$, i.e., if there is such a π—including s—such that for every $i \in I'$, $\pi \subseteq |E(x_n) \to \psi|^{\mathcal{M}'}f[i/n]$. Now, between the basis step of this proof, the identity of I and I', the induction hypothesis, and the foregoing argument concerning $\to$, for each $i \in I$, $|E(x_n) \to \psi|^{\mathcal{M}'}f[i/n]\restriction S = |E(x_n) \to \psi|^{\mathcal{M}}f[i/n]$. So if $0 \notin \pi$, this is immediately equivalent to $\pi \subseteq \sqcap_{i \in I}|E(x_n) \to \psi|^{\mathcal{M}}f[i/n]$. If $0 \in \pi$, then $\pi\restriction S \in Prop$ and $\pi\restriction S \in Prop'$, whence this is equivalent to $\pi\restriction S \subseteq \sqcap_{i \in I}|E(x_n) \to \psi|^{\mathcal{M}}f[i/n]$. Either way, there exists a $\pi' \in Prop$ witnessing the fact that $s \in \sqcap_{i \in I}|E(x_n) \to \psi|^{\mathcal{M}}f[i/n]$, i.e., $s \in |\forall x_n \psi|^{\mathcal{M}}f$. $\square$

Corollary 3 *The normalization $\mathcal{M}'$ of a model $\mathcal{M}$ is itself a model.*

Proof. Suppose that $\mathcal{M}$ is a model in the sense of Definition 3. Then for every formula φ and assignment f, $|\varphi|^{\mathcal{M}} f \in Prop$. But by Lemma 18, $|\varphi|^{\mathcal{M}'} f$ is either $|\varphi|^{\mathcal{M}} f$ itself (which is a member of $Prop'$ by construction) or $|\varphi|^{\mathcal{M}} f \cup \{0\}$ (which, too, is a member of $Prop'$). Thus, $\mathcal{M}'$ meets the criteria described in Definition 3. □

This establishes that for any model for Mares' **QR**, the method of normalization is adequate for our purposes.

3.2 γ-admissibility

Now, we are able to reap the fruits of the foregoing labor to establish that the rule γ is admissible. First, we observe that this work dovetails with the soundness and completeness of Mares' system with respect to the semantics to show that it is sound and complete with respect to the class of normal models.

Theorem 19 *For every sentence φ, φ is a theorem of Mares' varying-domain* **QR** *iff φ is valid in every normal frame.*

Proof. Every normal varying-domain **QR** model is *a fortiori* a varying-domain **QR** model, whence the left-to-right direction follows from soundness and completeness. For right-to-left, we prove the contrapositive. Suppose that φ is not a theorem of Mares' system. By completeness, one can pick a model $\mathcal{M}$ (not necessarily normal), assignment f, and a world $o \in N$ such that $o \notin |\varphi|^{\mathcal{M}} f$. But then one can construct the normalization of $\mathcal{M}$ at 0 for o and guarantee that $0 \notin |\varphi|^{\mathcal{M}'} f$, establishing that φ is not valid in every normal frame. □

Having shown this, we are now in a position to prove the desired result:

Theorem 20 γ *is admissible in Mares' varying-domain* **QR**

Proof. Suppose that φ and $\neg\varphi \vee \psi$ are theorems of the system. Then take a normal model $\mathcal{M}$ including a normal point $0 \in N$. For every f, $0 \in |\varphi|^{\mathcal{M}} f$ and $0 \in |\neg\varphi \vee \psi|^{\mathcal{M}} f$. Because $0 = 0^*$, that $0 \in |\varphi|^{\mathcal{M}} f$ means that $0^* \in |\varphi|^{\mathcal{M}} f$. But because $C00^*$, this means that there is a $t \in S$ such that $C0t$ and $t \in |\varphi|^{\mathcal{M}} f$, i.e., $0 \notin |\neg\varphi|^{\mathcal{M}} f$. But since $0 \in |\neg\varphi \vee \psi|^{\mathcal{M}} f$, this requires that $0 \in |\psi|^{\mathcal{M}} f$. As the normalization is arbitrary, ψ is valid and by completeness, ψ is a theorem. □

4 Concluding remarks

In the foregoing, we have made ample use of the strong assumptions that accompany a first-order logic whose propositional basis is **R**. Mares' technology governing varying domains, however, easily can be adapted to extend propositional relevant logics intermediate between **G** (**B** $+ \varphi \vee \neg\varphi$) and **R**. (*N.b.* that the technique of normalized models requires excluded middle in order to succeed.) Whether γ admissibility holds for a varying-domain first-order version of **DK**, for example, is left open; we anticipate, however, that the foregoing work lays much of the necessary foundations from which to set off on such expeditions.

A natural next step is to examine the extension of this method to a wide range of variable domain relevant logics based in relational semantics, and then to their modal extensions. We have shown here that the technique of normalization is able to accommodate several expansions to the commonly-encountered relevant logic idioms, namely, the *existence predicate*, the *compatibility relation*, and *varying-domain quantifiers*. We can view other work in this tradition through a similar lens, *e.g.*, interpret (Seki, 2011a) and (Ferenz & Ferguson, 2023) as demonstrating the robustness of the technique in its ability to incorporate *modalities* (in the first case) and *constant domain quantifiers*, *fusion*, and the *Ackermann constant* (in the latter). Future work should investigate how the technique fares when several of these devices are adopted in the same logic.

References

Ackermann, W. (1956). Begründung einer Strengen Implikation. *The Journal of Symbolic Logic*, *21*, 113–128.

Dunn, J. M. (1993). Star and perp: Two treatments of negation. *Philosophical Perspectives*, *7*, 331-357.

Ferenz, N., & Ferguson, T. M. (2023). γ-admissibility in first-order relevant logics: Proof using normal models in the Mares-Goldblatt setting. *Review of Symbolic Logic*. (To appear)

Goldblatt, R., & Hodkinson, I. (2009). Commutativity of quantifiers in varying-domain Kripke models. In D. Makinson, J. Malinowski, & H. Wansing (Eds.), *Towards Mathematical Philosophy* (pp. 9–30). Dordrecht: Springer.

Mares, E. D. (2009). General information in relevant logic. *Synthese*, *167*(2), 421–440.

Mares, E. D., & Goldblatt, R. (2006). An alternative semantics for quantified relevant logic. *The Journal of Symbolic Logic*, *71*, 163–187.

Routley, R., & Meyer, R. K. (1972). The semantics of entailment (2). *The Journal of Philosophical logic*, *1*, 53–73.

Routley, R., Plumwood, V., Meyer, R. K., & Brady, R. T. (1982). *Relevant Logics and Their Rivals: Part 1 The Basic Philosophical and Semantical Theory*. Atascadero, CA: Ridgewood.

Seki, T. (2011a). The γ-admissibility of relevant modal logics I — the method of normal models. *Studia Logica*, *97*, 199–231.

Seki, T. (2011b). The γ-admissibility of relevant modal logics II — the method using metavaluations. *Studia Logica*, *97*, 351–383.

Nicholas Ferenz
Czech Academy of Sciences, Institute of Computer Science
The Czech Republic
E-mail: ferenz@cs.cas.cz

Thomas M. Ferguson
Rensselaer Polytechnic Institute, Department of Cognitive Science
United States
E-mail: tferguson@gradcenter.cuny.edu

A Uniform Approach to Variable-Sharing in Relevant Logics

THOMAS M. FERGUSON[1] AND JITKA KADLEČÍKOVÁ[2]

Abstract: In this paper, we offer a uniform approach to the myriad variable-sharing properties characteristic of many relevant logics. We take cues from literature on the theory of topic and topic-sensitive logics in order to argue that the topic-theoretic contribution an atomic subformula makes to a complex is mediated by the intensional operators in which it appears, including (in the case of intensional conditionals) the chirality of the appearance. We introduce the notion of a *provenance* encapsulating the total intensional information concerning an atom's contribution to the overall topic of a complex and align a wide range of variable-sharing properties with equivalence relations over the set of provenances. This allows us to argue that differing variable-sharing properties correspond to differing degrees of intensional lossiness with respect to topic and argue that *lossless lericone* relevance holding of BM is in many ways the maximal natural variable-sharing property.

Keywords: variable-sharing property, relevant logic, theory of topic

1 Introduction

In several places (e.g. (Ferguson, 2023a)) we have argued that intensional connectives are *topic-theoretically transformative*, *i.e.*, the topics of sentences whose main operator is intensional need not be the mere mereological fusion of the topics of its subsentences. *E.g.*, where $\to$ is an intensional conditional, the topic of $p \to q$ might not be determined by the sum of the topics of p and q, but rather by a function applied to both topics, where a new topic emerges on top of the topics of the components. A number of examples are discussed at length in (Ferguson, 2023a).

---

[1] The comments of participants at Logica 2023 and an anonymous referee were particularly helpful to this paper, along with discussions with Shawn Standefer, Nicholas Ferenz, Andrew Tedder, and Shay Logan.

[2] This paper was supported by the Ministry of Education and Science of the Czech Republic granted to Palacký University, Olomouc (IGA_FF_2023_017).

Many works (*e.g.*, (Mares, 2022), (Tennant, 2017), and (Humberstone, 2011)) draw analogies between relevant logics' characteristic *variable-sharing properties* and relationships between subformulas' subject-matter or topic. The entrenchment of such analogies suggest that developments in the theory of topic ought to be fruitful tools in the analysis of relevant logics, a theme recently echoed by Øgaard:

> a better way of obtaining an intuitive motivation for the theory of rigorous implication... is by appealing to the notion of a topic or subject-matter. It is because p and q can be taken to express propositions about different topics that the relation of rigorous implication cannot obtain between $p \to p$ and $q \to q$. (Øgaard, 2023, p. 176)

Such insights, in tandem with a growing recognition of the importance of variable-sharing to the notion of relevance (as in, *e.g.*, Standefer's (Standefer, 2024b)), allow us to view the variable-sharing criterion of relevant logics through the lens of the theory of topic. As pointed out at length in, e.g., (Ferguson, 2024) and (Ferguson & Logan, 2023), various refinements to the variable-sharing criterion allow for a more fine-grained analysis of different degrees of relevance. However, an analogous opportunity for refinement arises on topic-theoretic grounds.

Considering the usefulness of both the different refinements of *variable-sharing* and their topic-theoretic interpretations further motivates the need for a uniform approach to the notion of relevance. Given that relevant logics differ with respect to the degree to which information is used to assess topic overlap, we can identify the quantity of information available with the granularity of equivalence relations. An increase in information, in general, corresponds to an ability to more and more finely discriminate between states. Total ignorance, on the one hand, amounts to a total equivalence relation that identifies any two objects; total information, on the other, amounts to an equivalence relation that relates φ and ψ just in case they are identical. We make use of this insight by suggesting a scale of information availability corresponding to the different degrees of refinement of the *variable-sharing criterion*.

2 Intensional Provenance

An often-encountered concern in recent work on the theory of topic (*e.g.* (Berto, 2022)) is the degree to which certain connectives and operators are

A Uniform Approach to Variable-Sharing

topic-transparent, that is, make no contribution to the overall topic of any complex in which they appear as main connective or operator. (Ferguson & Logan, 2023) has argued that the tacit topic-theoretic assumptions underlying certain families of relevant logic assume the failure of two species of transparency.

For one, the thesis of *negation transparency*—that the topic of a sentence φ and its negation $\neg\varphi$ necessarily coincide—appears to be violated by topic-theoretic readings both of weak relevance and the strong containment property advocated by Richard Angell in (Angell, 1989). For two, the thesis of *intensional transparency*—that the topic of an intensional conditional $\varphi \to \psi$ is nothing beyond the simple mereological fusion of the topics of its subformulae—conflicts with the property of depth relevance common to Ross Brady's preferred systems advocated for in (Brady, 1984).

Consequently, for an occurrence of a subformula p in a formula φ, the topic-theoretic contribution p makes to the overall subject-matter of φ is determined by the sequence of applications of intensional connectives. We can therefore construct a sort of record that includes further details regarding the order of applications or even the question of whether a given subformula resided in the antecedent or the consequent of the intensional conditional, such as represented in (Ferguson & Logan, 2023) and (Ferguson, 2024) by the notion of *provenance* of a formula, which can be understood as

> a record of the history of applications of the intensional conditional to the subformula... from which one may recover a detailed account of the influences over the overall subject-matter. (Ferguson, 2024, p. 8)

Assessing the contribution that p makes to φ, then, should correspond to a record of the sequence of intensional connectives that were applied to it. Following this line of reasoning, we introduce an overarching machinery, which is capable of accounting for an arbitrary detail of information. Let us call this system the *intensional alphabet* since all versions of the variable-sharing criterion can be mapped onto a graded, discrete space resembling in function that of the alphabet.

The maximally informative story of the influence of intensional connectives over the contribution of an atom can be given by a sequence determined by the intensional connectives within which it is nested. Let N, L, and R represent occurrences of negation, the left side of a conditional, and the right side of a conditional, respectively. Then:

Definition 1 *The intensional alphabet* **I** *is defined over the set of letters* $\{\mathtt{N}, \mathtt{L}, \mathtt{R}\}$. *The language* $\mathcal{L}_{\mathbf{I}} = \mathbf{I}^*$ *is the Kleene closure of* **I**.

Then the record of an appearance $\dot{p}$ in a formula φ (*i.e.* $\pi(\varphi[\dot{p}])$) can be recursively defined through the following scheme:

Definition 2 *The provenance of an appearance of formula ψ in φ is defined:*

- If $\varphi = \psi$ then $\pi(\varphi[\psi]) = \langle \rangle$
- $\pi(\neg \varphi[\psi]) = \pi(\varphi[\psi]) \frown \mathtt{N}$
- $\pi(\varphi[\psi] \wedge \xi) = \pi(\xi \wedge \varphi[\psi]) = \pi(\varphi[\psi])$
- $\pi(\varphi[\psi] \vee \xi) = \pi(\xi \vee \varphi[\psi]) = \pi(\varphi[\psi])$
- $\pi(\varphi[\psi] \to \xi) = \pi(\varphi[\psi]) \frown \mathtt{L}$
- $\pi(\xi \to \varphi[\psi]) = \pi(\varphi[\psi]) \frown \mathtt{R}$

Each element of $\mathbf{I}^*$ encodes a possible (and, arguably, a maximally fine-grained) account of the influence of intensional connectives over the topical contribution of an atom. Not all this information will be universally viewed as useful, however. For example, the thesis of negation transparency—that the topic of p is identical to the topic of $\neg p$—is one in which the character $\mathtt{N}$ is trivial.

If variable-sharing corresponds to topic overlap then the degree to which intensional connectives mediate assignments of topic will impart conditions on relevance. Such positions correspond to different equivalence relations. *I.e.*, to endorse negation transparency is to believe that any two provenances differing only on the appearances of negation will not differ with respect to their influence over the topic-theoretic contribution made by an atom. But this is just to say that two provenances differing only with respect to negations are *equivalent*.

We are thus compelled to introduce a schematic notion of relevance that is tied to such equivalence relations:

Definition 3 *Let $\sim$ be an equivalence relation on $\mathbf{I}^*$. Then a logic* L *is $\sim$-relevant iff for every theorem $\varphi \to \psi$, there exists an atom p appearing in both φ and ψ with instances such that* $\pi(\varphi[\dot{p}]) \sim \pi(\psi[\ddot{p}])$

Indeed, as we will see, all of the various variable-sharing properties in the literature with which we are familiar enjoy such a characterization. The upshot is that different variable-sharing properties—and the logics that exhibit them—can be given a gloss in terms of their tacit theories about how intensional connectives influence topic assignments.

To begin, we will provide such an analysis and interpretation of the traditional variable-sharing properties. We will then move to more recent families of variable-sharing properties with even more discriminating accounts of the topic-theoretic influence of intensional connectives. We will conclude by investigating a property of *losslessness* discussed in (Ferguson & Logan, 2023) and the interpretation of the variable-sharing properties exhibiting losslessness.

3 Traditional VSPs

In this section, we track the different refinements of the variable-sharing property. Furthermore, we provide a unification in the form of a translation of these relevant logic properties into the intensional alphabet sketched above. *I.e.*, we show that every property can be expressed as an equivalence relation on $\mathbf{I}^*$.

Starting from *variable-sharing* and continuing through *strong variable-sharing*, *depth relevance*, and *strong depth relevance*, we show that all of these properties can be harnessed by one uniform notation of the equivalence relations sketched above. To begin, we will first introduce the variable-sharing criterion:

Definition 4 *A logic* L *enjoys the variable-sharing property if for every* L-*theorem* $\varphi \to \psi$ *there is some atom appearing in both* φ *and* ψ.

This well-known principle first introduced in Belnap's dissertation (Belnap, 1959) (and discussed at length in (Anderson & Belnap, 1975)) has been taken by many to be symptomatic of *topic overlap*, such as discussed *e.g.* in (Tennant, 2017) or recently (Berto, 2022) and (Ferguson, 2024). Making use of the novel notation of the intensional alphabet, the vanilla variable-sharing property can be rephrased in terms of an equivalence relation like so:

Definition 5 *The* trivial equivalence $\sim_{\mathsf{vsp}}$ *is defined s.t. for* $\sigma, \tau \in \mathbf{I}^*$,

$$\sigma \sim_{\mathsf{vsp}} \tau$$

$\sim_{\mathsf{vsp}}$ represents absolute coarseness. The influence of intensional connectives on determining the topic-theoretic contribution made by an atom is nullified. We might think of this as corresponding to the position embracing both negation and intensional transparency, *i.e.*, the contribution that p makes in any formula is the same as the topic of p. More generally, then, the following can be noted:

Observation 1 *A logic has the variable-sharing property iff it is $\sim_{\mathsf{vsp}}$-relevant.*

3.1 Equivalences in Virtue of Sign

The traditional representative of what we call equivalence in virtue of sign is *strong variable-sharing*. Importantly, however, strong variable-sharing is merely a specific example of a universal phenomenon of sign equivalence. To provide the basic steps to generalize this relation, we proceed to summarize the notions of *sign* and *strong variable-sharing*:

Definition 6 *The* sign *of an occurrence of ψ appearing in φ is defined:*

- *ψ appears positively in ψ*
- *if ψ appears positively (resp, negatively) in φ, then:*
    - *ψ appears negatively (positively) in $\neg \varphi$*
    - *ψ appears positively (negatively) in $\varphi \wedge \xi$ and $\xi \wedge \varphi$*
    - *ψ appears positively (negatively) in $\varphi \vee \xi$ and $\xi \vee \varphi$*
    - *ψ appears negatively (positively) in $\varphi \rightarrow \xi$*
    - *ψ appears positively (negatively) in $\xi \rightarrow \varphi$*

Definition 7 *A sentence $\varphi \rightarrow \psi$ exhibits the* strong variable-sharing *property if some atomic variable appears in both φ and ψ with the same sign.*

This leads to the definition of the refined property in the case of a logic:

Definition 8 *A logic L enjoys the strong variable-sharing property if every L-theorem $\varphi \rightarrow \psi$ exhibits strong variable-sharing.*

Alternatively, however, we can generalize this property, which allows for a uniform treatment of *any* possible equivalences in virtue of *sign*. We can generalize the notion of the number of occurrences, which will allow us to describe sign equivalence for any arbitrary number of embedded signs.

Definition 9 *Let X be a set of strings. Then $\#(X, \sigma)$ is the number of occurrences of elements of X in the string σ*

A Uniform Approach to Variable-Sharing

This machinery allows us to define the sign equivalence relation:

Definition 10 *The* sign equivalence $\sim_{\text{svsp}}$ *is defined s.t. for* $\sigma, \tau \in \mathbf{I}^*$,

$$\sigma \sim_{\text{svsp}} \tau \text{ iff } \#(\{\texttt{N}, \texttt{L}\}, \tau) = \#(\{\texttt{N}, \texttt{L}\}, \sigma)$$

Looking back at the original strong variable-sharing property, we might define it retrospectively as an example of sign equivalence:

Observation 2 *A logic has the strong variable-sharing property iff it is* $\sim_{\text{svsp}}$-*relevant.*

Notice that Definition 10 provides an elegant notation for the strong variable-sharing property. This notation, moreover, does not require us to define sign in a case-by-case fashion.

To summarize, this move allows us to subsume vanilla variable-sharing and strong variable-sharing under a uniform theory. Analogously, below, we show that this holds about any traditional variable-sharing property.

3.2 Equivalences in Virtue of Depth

Similarly, we can expose the same redefinition in the case of *depth relevance*. Acknowledging the transformative nature of the intensional conditional, *depth* is understood as the nesting of a subformula within a certain number of conditionals, such as introduced in (Brady, 1984).

Definition 11 *For an occurrence of* ψ *appearing in* φ, depth *is defined:*

- ψ *appears at depth* 0 *in* ψ
- *if* ψ *appears at depth* n *in* φ, *then:*
  - ψ *appears at depth* n *in* $\neg\varphi$
  - ψ *appears at depth* n *in* $\varphi \wedge \xi$ *and* $\xi \wedge \varphi$
  - ψ *appears at depth* n *in* $\varphi \vee \xi$ *and* $\xi \vee \varphi$
  - ψ *appears at depth* $n + 1$ *in* $\varphi \to \xi$ *and* $\xi \to \varphi$

Essentially, the depth of ψ in φ is a discrete measure of the degree to which intensional conditionals have operated on the topic of ψ within complex φ.

Definition 12 *A sentence* $\varphi \to \psi$ *exhibits the* depth relevance property *if some atomic variable appears in both* φ *and* ψ *at the same depth.*

This allows us to define depth relevance of a logic:

Definition 13 *A logic* L *enjoys depth relevance if every* L*-theorem* $\varphi \to \psi$ *exhibits the depth relevance property.*

Brady motivates depth relevance in virtue of *meaning containment*; (Standefer, 2024a) includes a particularly nice discussion of the link between depth relevance and meaning containment.

Alternatively, we can redefine depth relevance in terms of equivalence relations labeled *depth equality*:

Definition 14 *The* depth equality $\sim_{\mathbf{dr}}$ *is defined s.t. for* $\sigma, \tau \in \mathbf{I}^*$,

$$\sigma \sim_{\mathbf{dr}} \tau \text{ iff } \#(\{\text{L}, \text{R}\}, \tau) = \#(\{\text{L}, \text{R}\}, \sigma)$$

Again, we can see that the depth relevance property is definable using the same mechanisms as before, without having to define it case by case for each connective, as we were forced to do in Definition 11. In the case of logics rather than formulae, an analogous observation holds:

Observation 3 *A logic has depth relevance iff it is* $\sim_{\mathbf{dr}}$-*relevant.*

3.3 Strong Depth Relevance

Logan (2021) introduces the notion of *strong depth relevance* which acknowledges the transformative nature of both negation and the intensional conditional. As a result, we get a system consisting of both equivalences in virtue of sign and equivalences in virtue of depth.

Definition 15 *A sentence* $\varphi \to \psi$ *exhibits the* strong depth relevance *property if some atomic variable appears in both* φ *and* ψ *at the same depth and with the same sign.*

We will describe two formulae φ and ψ as sharing strong depth relevance if the conditional $\varphi \to \psi$ exhibits the property.

Definition 16 *A logic* L *enjoys strong depth relevance if every* L*-theorem* $\varphi \to \psi$ *exhibits strong depth relevance.*

Definition 17 *The* strong depth equivalence $\sim_{\mathbf{sdr}}$ *is defined for* $\sigma, \tau \in \mathbf{I}^*$:

$$\sigma \sim_{\mathbf{sdr}} \tau \text{ iff } \sigma \sim_{\mathbf{svsp}} \tau \text{ and } \sigma \sim_{\mathbf{dr}} \tau$$

Observation 4 *A logic has strong depth relevance iff it is* $\sim_{\mathbf{sdr}}$-*relevant*

The strong variable-sharing property and depth relevance are the traditional mainstays of variable-sharing properties. Although its introduction in the literature is more recent, Definition 17 and Observation 4 show that Logan's strong depth relevance of (Logan, 2021) involves no more discrimination in intensional information than the conjunction of these two properties.

4 Fine-Grained VSPs

Beyond these variable-sharing properties, however, a host of upstarts have been introduced that are interesting for their more refined ability to differentiate between the roles that particular variables play in the determination of the topics expected to overlap. This refinement, we suggest, can be explicitly—and uniformly—viewed as these systems' encoding finer mechanisms to discern between provenances. In this section, we will examine the space of these more recent properties and characterize them in terms of corresponding equivalence classes on $\mathbf{I}^*$, several of which make use of similar provenance-like objects defined over other alphabets.

4.1 Genealogical Relevance

The recent (Ferguson, 2024) considers examples to emphasize not only that intensional conditionals—like those employed in relevant logics—exert a topic-transformative effect on the subformulae appearing in them but that the matter of whether a sentence appears in the antecedent or consequent is also topic-theoretically salient. There are, for example, clear cases in which a conditional $p \to q$ seems to have a markedly different topic than the converse $q \to p$.

Although our language $\mathbf{I}^*$ allows records of negations, in (Ferguson, 2024), a *genealogical* variable-sharing property focusing only on chirality is shown to hold of the containment logic CA/PAI introduced in (Ferguson, 2023a).[3] That paper's notion of a genealogy is essentially a flattened version of the present provenances, defined over the following language:

Definition 18 *The* strictly chiral language $\mathbf{Ch}^*$ *is defined over the chiral alphabet* $\{\mathrm{L}, \mathrm{R}\}$.

And the assignment of genealogy is defined recursively over $\mathbf{Ch}^*$ as follows:

---

[3] CA/PAI—being related to William Parry's PAI of (Parry, 1968)—is not a "mainstream" relevant logic. Nevertheless, as the following will make clear, the genealogical variable-sharing property can be seen to hold of *e.g.* the logic B of (Routley, Plumwood, Meyer, & Brady, 1982).

Thomas M. Ferguson and Jitka Kadlečíková

Definition 19 *The genealogy of an appearance of a formula ψ in a formula ξ—$\iota(\xi[\psi])$—is defined as follows:*

- If $\varphi = \psi$ then $\iota(\varphi[\psi]) = \langle\rangle$
- $\iota(\neg\varphi[\psi]) = \iota(\varphi[\psi])$
- $\iota(\varphi[\psi] \wedge \xi) = \iota(\xi \wedge \varphi[\psi]) = \iota(\varphi[\psi])$
- $\iota(\varphi[\psi] \vee \xi) = \iota(\xi \vee \varphi[\psi]) = \iota(\varphi[\psi])$
- $\iota(\varphi[\psi] \to \xi) = \iota(\varphi[\psi]) \frown \mathrm{L}$
- $\iota(\xi \to \varphi[\psi]) = \iota(\varphi[\psi]) \frown \mathrm{R}$

Definition 20 *A logic has the* genealogical variable-sharing property *if for all theorems of the form $\varphi \to \psi$, there exists an atom p with occurrences in φ and ψ such that $\iota(\varphi[\dot{p}]) = \iota(\psi[\dot{p}])$.*

In order to characterize genealogical variable-sharing in our uniform framework, we introduce first some notation:

Definition 21 *The restriction of a string to a set of letters X—$\sigma\upharpoonright X$—is the restriction of σ to only letters appearing in X.*

This allows us to provide an equivalence relation on provenances that characterizes the genealogical VSP:

Definition 22 *The* genealogical equivalence $\sim_{\mathrm{gen}}$ *is defined for $\sigma, \tau \in \mathbf{I}^*$:*

$$\sigma \sim_{\mathrm{gen}} \tau \text{ iff } \sigma\upharpoonright\{\mathrm{L},\mathrm{R}\} = \tau\upharpoonright\{\mathrm{L},\mathrm{R}\}$$

Which permits the following observation:

Observation 5 *A logic has genealogical variable-sharing iff it is $\sim_{\mathrm{gen}}$-relevant*

The nature of $\sim_{\mathrm{gen}}$ is such that chirality is acknowledged as a necessary datapoint with respect to uncovering the topic-theoretic influence of intensional connectives but negation is not. The underlying thesis thus aligns with an embrace of *negation transparency* while assuming an even stronger thesis concerning conditionals than is assumed by Brady's depth relevance, *i.e.*, a position in which not only the conditional but also the argument place determines topic assignments.

A Uniform Approach to Variable-Sharing

4.2 cn-Relevance

Although the definition of $\mathbf{I}^*$ allows for capturing the chirality of a subformula's appearance within a conditional, that the position in which a sentence appears within the scope of an intensional is topic-theoretically influential is debatable. One might find using the letters L and R to track chirality to be superfluous, maintaining that a C (for "conditional") captures all the necessary and salient contributions made by a conditional.

Ferguson and Logan (2023) argue that a property of *cn-relevance* captures such a position and shows that it holds of the logic B of (Routley et al., 1982). But such a position—conflating L and R while acknowledging the influence of negations—does not collapse into strong depth relevance. cn-relevance, like genealogical relevance, provides a sort of refinement of Brady's intuition. Formally, (Ferguson & Logan, 2023) invokes an object not dissimilar to our provenances.

Definition 23 *The* cn *language* $\mathbf{Cn}^*$ *is defined over the cn-alphabet* $\{\mathtt{C}, \mathtt{N}\}$.

And the assignment of genealogy is defined recursively over $\mathbf{Cn}^*$ as follows:

Definition 24 *The cn-signature of an appearance of a formula ψ in a formula ξ—$\kappa(\xi[\psi])$—is defined as follows:*

- *If $\varphi = \psi$ then $\kappa(\varphi[\psi]) = \langle\rangle$*
- $\kappa(\neg\varphi[\psi]) = \kappa(\varphi[\psi]) \frown \mathtt{N}$
- $\kappa(\varphi[\psi] \wedge \xi) = \kappa(\xi \wedge \varphi[\psi]) = \kappa(\varphi[\psi])$
- $\kappa(\varphi[\psi] \vee \xi) = \kappa(\xi \vee \varphi[\psi]) = \kappa(\varphi[\psi])$
- $\kappa(\varphi[\psi] \to \xi) = \kappa(\xi \to \varphi[\psi]) = \kappa(\varphi[\psi]) \frown \mathtt{C}$

Ferguson and Logan (2023) introduce a notion of *similarity* between strings in order to characterize cn-relevance.

Definition 25 $\sigma \sim_{\mathrm{lrcn}'} \tau$ *iff there exist strings π and ξ such that:*

$$\sigma = \pi \frown \xi \text{ and } \tau = \pi \frown \mathtt{NN} \frown \xi$$

$\sim_{\mathrm{lrcn}}$ *is the reflexive, symmetric, and transitive closure of* $\sim_{\mathrm{lrcn}'}$

We say that two strings are *similar* if they are equivalent modulo $\sim_{\mathrm{lrcn}}$. *N.b.* that $\sim_{\mathrm{lrcn}}$ can be defined on the language $\mathbf{I}^*$ as well as on $\mathbf{Cn}^*$.

Again, we find connections to the theory of intensional topic emerge. $\sim_{lrcn}$ is an interesting relation from the perspective of *negation transparency*. It assumes that although negation might not be topic-transparent, it is *involutively transparent*, that is, even if the topic of φ and $\neg\varphi$ may differ, those of φ and $\neg\neg\varphi$ must coincide.

The following variable-sharing properties are introduced (along with a further property which will be discussed in Section 5.2). These are:

Definition 26 *A logic* L *enjoys the* cn-relevance property *if for every theorem* $\varphi \to \psi$ *there exist occurrences* $\varphi[\dot{p}]$ *of p in* φ *and* $\psi[\dot{p}]$ *of p in* ψ *for which* $\kappa(\varphi[\dot{p}])$ *and* $\kappa(\psi[\dot{p}])$ *are similar.*

Definition 27 *A logic* L *enjoys the* strong cn-relevance property *if for every theorem* $\varphi \to \psi$ *there exist occurrences* $\varphi[\dot{p}]$ *of p in* φ *and* $\psi[\dot{p}]$ *of p in* ψ *for which:*

- $\varphi[\dot{p}]$ *and* $\psi[\dot{p}]$ *have identical sign, and*
- $\kappa(\varphi[\dot{p}])$ *and* $\kappa(\psi[\dot{p}])$ *are similar*

In order to bring these properties in line with our uniform treatment, let $\sigma[\mathtt{X} := \mathtt{Y}]$ be the string resulting in the replacement of each occurrence of X with an occurrence of Y. Then:

Definition 28 *The* cn equivalence $\sim_{cn}$ *is defined s.t. for* $\sigma, \tau \in \mathbf{I}^*$,

$$\sigma \sim_{cn} \tau \text{ iff } \sigma[\{\mathtt{L}, \mathtt{R}\} := \mathtt{C}] \sim_{lrcn} \tau[\{\mathtt{L}, \mathtt{R}\} := \mathtt{C}]$$

Definition 29 *The* strong cn equivalence $\sim_{scn}$ *is defined as* $\sim_{svsp} \cap \sim_{cn}$

This allows us the following characterizations:

Observation 6 *A logic has the cn-relevance property iff it is* $\sim_{cn}$*-relevant and has the strong cn-relevance property if it is* $\sim_{scn}$*-relevant.*

4.3 Lericone Relevance

The final item among our fine-grained variable-sharing properties is so-called *lericone relevance*. Discussed in (Standefer, Logan, & Ferguson, n.d.), lericone (L − R − C − N) relevance is a type of variable-sharing property that takes advantage of nearly all the information encoded in our provenances.

Lericone relevance is characterized by an equivalence relation already introduced: that of *similarity*, the nomenclature for which—$\sim_{lrcn}$—was selected for these purposes.

Definition 30 *A logic* L *enjoys the* lericone relevance property *if for every theorem* $\varphi \to \psi$ *there exist occurrences* $\varphi[\dot{p}]$ *of p in* φ *and* $\psi[\ddot{p}]$ *of p in* ψ *for which* $\pi(\varphi[\dot{p}])$ *is similar to* $\pi(\psi[\ddot{p}])$.

This immediately establishes that lericone relevance is $\sim_{\text{lrcn}}$ relevance.

We can observe here that—in contrast to the themes of (Logan, 2021) and (Logan, 2022)—there is no corresponding "strong" notion of lericone relevance, that is, agreement in sign follows from similarity. Thus:

Observation 7 $\quad \sim_{\text{lrcn}} = (\sim_{\text{lrcn}} \cap \sim_{\text{svsp}})$

And consequently:

Corollary 1 *A logic is lericone relevant iff it is strongly lericone relevant.*

Lericone relevance avails itself of nearly all of the topic-theoretic information concerning the contributions of the intensional nesting of atoms. The only topic-theoretic coarseness reflected in the assumption that involutions of negation are topic-transparent. It is thus the most refined variable-sharing property encountered thus far. But there is a further refinement possible by emphasizing the loss of information due to this involutive transparency.

5 Losslessness and Negation Transparency

The provenance of an atom's appearance in a complex is an information-rich object; that not all information included therein is leveraged in a particular variable-sharing property means that the topic-theoretic information required by that property has some tolerance towards *lossiness*. Assessing genealogical relevance, for example, requires maximal information concerning the influence of the conditional on atoms' topic-theoretic contributions but conflates information concerning sign or negation.

We will now look at the opposite case, in which no lossiness concerning negation is tolerated before tying the pieces together to describe what we take to be the maximally lossless relevance property that can be articulated in our framework.

5.1 Modulus Relevance

Although (Ferguson, 2018) is devoted to studying the semantics of Sylvan's *mate function* in cases in which it is cyclic, the work introduces an infinite collection of variable-sharing properties as well. These variable-sharing properties align with the modulus of the cycle of the mate function itself, including a property characteristic of each logic whose logic has a mate function with modulus n:

Definition 31 *A logic has the cyclical variable-sharing property with modulus n if a formula $\varphi \to \psi$ is a theorem only if there exists an atom p and numbers j and k such that $j \equiv k \mod n$ for which p appears within the scope of j many negation signs in φ and k many negation signs in ψ.*

This infinite sequence of properties has a limiting case that characterizes the calculus at the intersection of all of the logics introduced in (Ferguson, 2018):

Definition 32 *A logic has the limiting cyclical variable-sharing property if a formula $\varphi \to \psi$ is a theorem only if there exists an atom p appearing within the scope of the same number of negation signs in φ and in ψ.*

As acknowledged in (Ferguson & Logan, 2023), the above properties are worth acknowledging for two reasons. First, these properties bear witness to the overall utility of approaching variable-sharing recommended in this paper. Second, the limiting case is interesting from a topic-theoretic perspective and helps motivate the species of *lossless* variable-sharing properties introduced in (Ferguson & Logan, 2023). We first treat the first of these reasons by showing how they correspond to particular equivalence classes on $\mathbf{I}^*$. Consider the following equivalence relation:

Definition 33 *The equivalence relation $\sim_{\mod_n}$ is defined s.t. for $\sigma, \tau \in \mathbf{I}^*$,*

$$\sigma \sim_{\mod_n} \tau \text{ iff } \#(\{\mathtt{N}\}, \tau) \equiv \#(\{\mathtt{N}\}, \sigma) \mod n$$

Observation 8 *A logic has the cyclical variable-sharing property with modulus n iff it is $\sim_{\mod_n}$-relevant*

Definition 34 *The equivalence relation $\sim_{\mod_\omega}$ is defined s.t. for $\sigma, \tau \in \mathbf{I}^*$,*

$$\sigma \sim_{\mod_\omega} \tau \text{ iff } \#(\{\mathtt{N}\}, \tau) = \#(\{\mathtt{N}\}, \sigma)$$

Observation 9 *A logic has the limiting cyclical variable-sharing property iff it is $\sim_{\mod_\omega}$-relevant*

In the first-degree case discussed in (Ferguson, 2018), $\sim_{\mod_\omega}$ represents an insistence of total losslessness concerning negation in order to ensure topic overlap of two variables' contributions. This resists a sort of *involutive transparency* that even opponents of negation transparency (like *e.g.* Angell in (Angell, 1989)) in general accept. (But see (Ferguson, 2023b).)

We have already encountered losslessness with respect to conditionals; in the next section, we put these pieces together to examine what we take to be the maximally lossless relevance property as far as intensional contributions are concerned.

5.2 Lossless Versions of Fine-Grained VSPs

In (Ferguson & Logan, 2023), a *lossless* version of cn-relevance was introduced in which the topic-theoretic information concerning negation's effects that are lost upon involution are preserved.

Definition 35 *A logic* L *enjoys the* lossless cn-relevance property *if for every theorem* $\varphi \to \psi$ *there exist occurrences* $\varphi[\dot{p}]$ *of p in* φ *and* $\psi[\dot{p}]$ *of p in* ψ *for which* $\kappa(\varphi[\dot{p}]) = \kappa(\psi[\dot{p}])$.

This is shown to hold of the Priest-Sylvan system BM (of (Priest & Sylvan, 1992)) whose semantics drop the involutive nature of the Routley star.

Definition 36 *The* lossless cn equivalence $\sim_{\text{lcn}}$ *is defined s.t. for* $\sigma, \tau \in \mathbf{I}^*$,

$$\sigma \sim_{\text{cn}} \tau \text{ iff } \sigma[\{\mathtt{L},\mathtt{R}\} := \mathtt{C}] = \tau[\{\mathtt{L},\mathtt{R}\} := \mathtt{C}]$$

This allows us the following characterizations:

Observation 10 *A logic has the lossless cn-relevance property iff it is* $\sim_{\text{lcn}}$-*relevant.*

We can note, also, that there is a distinct lossless *strong* notion of relevance had by defining:

Definition 37 *The* lossless strong cn equivalence $\sim_{\text{lscn}}$ *is defined as* $(\sim_{\text{svsp}} \cap \sim_{\text{lcn}})$

The reader can observe that $\sim_{\text{lscn}}$-relevance is distinct from $\sim_{\text{lcn}}$-relevance by considering the formulae $p \to q$ and $q \to p$. Finally, in (Standefer et al., n.d.), the feature of lericone relevance can be associated with a property of *lossless lericone relevance* which, likewise, can be shown to hold of BM.

Definition 38 *A logic* L *enjoys* lossless lericone relevance *if for every theorem* $\varphi \to \psi$ *there exist occurrences* $\varphi[\dot{p}]$ *of p in* φ *and* $\psi[\dot{p}]$ *of p in* ψ *for which* $\pi(\varphi[\dot{p}]) = \pi(\psi[\dot{p}])$.

Alternatively, however, we can define the most fine-grained equivalence relation on $\mathbf{I}^*$ by defining an equivalence relation $\sigma \sim_{\text{llrcn}} \tau$ as identity itself and characterize lossless lericone relevance in this way:

Observation 11 *A logic has the lossless lericone relevance property iff it is $\sim_{\text{llcn}}$-relevant.*

One further point can be made relating the lossless versions of these fine-grained variable-sharing properties and the modulus relevance properties.

Observation 12 *The lossless relevance properties are also $\sim_{\text{mod}_\omega}$-relevant.*

6 Discussion and Concluding Remarks

Put together, both the traditional variable-sharing properties and their more recently developed refined variants can be represented in a diagram, where they are ordered by the degree of their *lossiness* (from top to bottom), or, alternatively, by the degree of their *losslessness* (from bottom to top), like in Figure 1. Note that the inclusions of equivalence relations are represented by arrows, with the arrow pointing towards coarsenings.

If $\sim_{\text{vsp}}$ reveals the vanilla variable-sharing property to be absolutely blunt with respect to the topic-theoretic contributions of intensional operators, then lossless lericone relevance is maximally fine-grained. Under the assumption of extensional transparency, any further discriminations one could make will be topic-theoretically immaterial.

Despite the apparent exhaustive nature of this diagram, several questions can still be raised regarding the borderline cases. First, we might ask whether this is inevitable, *i.e.*, whether we could not introduce any further refinements down the line. *E.g.*, one could elaborate to add characters to the intensional alphabet $\mathbf{I}$ for conjunction and disjunction (or even the chiral versions of these) to track the potential topic-theoretic influence of these extensional connectives. Certainly, intensional relevant connectives like fusion (or, indeed, any connectives that satisfy Standefer's criteria described in (Standefer, 2022)) seem to warrant representation in $\mathbf{I}$.

Philosophically, there is a case to be made that given some pragmatic considerations, the topics of p and $p \wedge p$ might differ, just as, for instance, the topics of $p \wedge q$ and $q \wedge p$. In addition, there is no *a priori* argument against having even more fine-grained machinery extend our semantics to incorporate such cases if they are well motivated. Our theory, therefore,

A Uniform Approach to Variable-Sharing

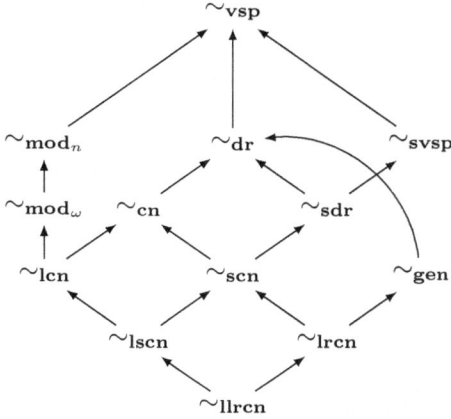

Figure 1: Inclusions Between Equivalence Relations on **I***

could be expanded to incorporate further refinements, such as the potential extensional topic-transformative connectives. Even though we are not explicitly developing these ideas, there is nothing inherently preventing an expansion in that direction.

Second, we might ask whether the vanilla variable-sharing property truly corresponds to topic-theoretically maximal coarseness. If we wanted to generalize the notion of relevance in this framework, rather than merely relying on variable-sharing as its marker, one might ask how we are utilizing the full scale of relevance if we are not considering *minimal relevance*, or simply the lack thereof. In an important sense, there are coarser equivalence relations than the variable-sharing property: namely, no topic overlap at all.

Finally, there is an interesting additional dimension to the matter of "coarseness" worth investigating. (Leach-Krouse, Logan, & Worley, 2024) has proceeded to show that there is a maximal fragment of R that satisfies depth relevance (which turns out to be recursively axiomatizable). A natural question to ask is which of the above variable-sharing properties have maximal witnesses, which of these are recursively axiomatizable, and—if there exist some such systems—whether there is any pattern concerning which variable-sharing properties have such maximal systems.

In other words, our approach does not allow for expressing any equivalence relations that do not involve any variable-sharing property, such as logical systems that are *not* topic-sensitive. However, a truly uniform account of relevance would be capable of treating these cases as well, presupposing that absolute coarseness corresponds to no relevance, and thus, no topic overlap whatsoever. There are, therefore, at least two controversial points to be considered in future work.

References

Anderson, A. R., & Belnap, N. D., Jr. (1975). *Entailment: The Logic of Relevance and Necessity* (Vol. I). Princeton, NJ: Princeton University Press.

Angell, R. B. (1989). Deducibility, entailment and analytic containment. In J. Norman & R. Sylvan (Eds.), *Directions in Relevant Logic* (pp. 119–143). Boston, MA: Kluwer Academic Publishers.

Belnap, N. D. J. (1959). *The Formalization of Entailment* (Unpublished doctoral dissertation). Yale University, New Haven.

Berto, F. (2022). *Topics of Thought*. Oxford: Oxford University Press.

Brady, R. T. (1984). Depth relevance of some paraconsistent logics. *Studia Logica*, *43*(1–2), 63–73.

Ferguson, T. M. (2018). Parity, relevance, and gentle explosiveness in the context of Sylvan's mate function. *Australasian Journal of Logic*, *15*(2), 381–406.

Ferguson, T. M. (2023a). Subject-matter and intensional operators I: Conditional-agnostic analytic implication. *Philosophical Studies*, *180*(7), 1849–1879.

Ferguson, T. M. (2023b). Subject-matter and intensional operators II: Applications to the theory of topic-sensitive intentional modals. *Journal of Philosophical Logic*, *52*(6), 1673–1701.

Ferguson, T. M. (2024). A topic-theoretic perspective on variable-sharing (from the black sheep of the family). In I. Sedlár, S. Standefer, & A. Tedder (Eds.), *New Directions in Relevant Logic*. Cham: Springer. (To appear)

Ferguson, T. M., & Logan, S. A. (2023). Topic transparency and variable sharing in weak relevant logics. *Erkenntnis*. (To appear)

Humberstone, L. (2011). *The Connectives*. Cambridge, MA: MIT Press.

Leach-Krouse, G., Logan, S. A., & Worley, B. (2024). Logic in the deep end. *Analysis*. (To appear)

Logan, S. A. (2021). Strong depth relevance. *Australasian Journal of Logic*, *18*(6), 645–656.

Logan, S. A. (2022). Depth relevance and hyperformalism. *Journal of Philosophical Logic*, *51*(4), 721–737.

Mares, E. (2022). Relevance logic. In E. N. Zalta & U. Nodelman (Eds.), *The Stanford Encyclopedia of Philosophy* (Fall 2022 ed.).

Øgaard, T. F. (2023). Relevance through topical unconnectedness: Ackermann and Plumwood's motivational ideas on entailment. *Australasian Journal of Logic*, *20*(2), 154–187.

Parry, W. T. (1968). The logic of C. I. Lewis. In P. A. Schilpp (Ed.), *The Philosophy of C. I. Lewis* (pp. 115–154). La Salle, IL: Open Court.

Priest, G., & Sylvan, R. (1992). Simplified semantics for basic relevant logics. *Journal of Philosophical Logic*, *21*(2), 217–232.

Routley, R., Plumwood, V., Meyer, R. K., & Brady, R. (1982). *Relevant Logics and their Rivals* (Vol. 1). Atascadero, CA: Ridgeview Publishing.

Standefer, S. (2022). What is a relevant connective? *Journal of Philosophical Logic*, *51*(4), 919–950.

Standefer, S. (2024a). Routes to relevance: Philosophies of relevant logics. *Philosophy Compass*, *19*(2), 1–18. (e12965)

Standefer, S. (2024b). Variable-sharing as relevance. In I. Sedlár, S. Standefer, & A. Tedder (Eds.), *New Directions in Relevant Logic*. Cham: Springer. (To appear)

Standefer, S., Logan, S. A., & Ferguson, T. M. (n.d.). *Topics, non-uniform substitutions, and variable sharing*. (Under review)

Tennant, N. (2017). *Core Logic*. Oxford: Oxford University Press.

Thomas M. Ferguson
Rensselaer Polytechnic Institute, Department of Cognitive Science
United States
E-mail: `tferguson@gradcenter.cuny.edu`

Jitka Kadlečíková
Rensselaer Polytechnic Institute, Department of Cognitive Science
United States
Palacký University Olomouc, Department of Philosophy
Czech Republic
E-mail: `kadlej@rpi.edu`

Reconsidering Identity

RENÉ GAZZARI[1]

Abstract: We address the problem of whether equality should be considered as a logical symbol. The investigation is based on the proof-theoretic analysis of a new kind of introduction and elimination rules for equality, namely the proper term rules of Natural Calculation (proposed by Gazzari, 2021). We argue that these rules are harmonious, as they satisfy an inversion principle à la Prawitz, and that therefore the equality symbol is logical. The latter motivates an extension of the BHK interpretation of logic. The proposed interpretation of equality is used to represent the term rules within standard Natural Deduction, which clarifies the nature of the standard identity rules and the reasons why none of the standard rules should be considered as the introduction of the equality.

Keywords: equality, identity, harmony, inversion principle, proof-theoretic semantics

1 Introduction

Gentzen (1934) observes in his seminal investigations[2] a remarkable harmony between the introduction and the elimination rules for the logical operators in his calculus of Natural Deduction (ND). Moreover, he considers the introduction rules as definitions of the associated operators and expects that, in the final analysis, the elimination rules can be determined by them.

Turning slightly the perspective, harmonious introduction and elimination rules became (at least form the perspective of proof-theoretic semantics) a criterion for the logicality of the symbols governed by these rules. In other words, it is claimed that harmony can be used, in the case of logical operators, to determine (or to justify) the elimination rules given the introduction rules.

---

[1] I am grateful to José Espírito Santo and Luís Pinto for their patient support during all stages of the preparation of this paper as well as to my anonymous reviewer for his kind and accurate comments on the previous version of this paper.

This work is funded by national funds through the Portuguese Foundation for Science and Technology (FCT) by the projects with references UIDB/00013/2020 and UIDP/00013/2020.

[2] There is an English translation of Gentzen's investigations by Szabo (Gentzen, 1969); for accuracy reasons, we prefer the German original.

René Gazzari

The equality symbol is usually considered as a logical symbol, the reflexivity rule (I_1) as its introduction rule and the congruence (formula) rule (I_5) as its elimination rule (the standard ND identity rules are stated in figure 1). Unfortunately, (I_1) seems to be far too weak to be in harmony with (I_5).

$$\frac{}{t = t}\ (I_1) \quad ; \quad \frac{t = s}{s = t}\ (I_2) \quad ; \quad \frac{t = s \quad s = r}{t = r}\ (I_3)$$

$$\frac{t = s}{\mathtt{r}[t] = \mathtt{r}[s]}\ (I_4) \quad ; \quad \frac{\mathtt{A}[t] \quad t = s}{\mathtt{A}[s]}\ (I_5)$$

In the congruence rules, $\mathtt{r}$ and $\mathtt{A}$ are unary nominal forms (nominal forms are explained in remark 1); the terms t and s must be free for substitution in $\mathtt{A}$.

Figure 1: Standard identity rules of ND

State of the art. In order to solve this problem, Read (2004) suggests a stronger introduction rule obtained from the analysis of Leibniz's Identity formulated in second-order logic. One problem of his suggestion is his use of a predicate variable F in the first-order formulation of his introduction rule. In our conception of first-order logic, there are no predicate variables available. Using an "arbitrary" relation symbol F instead (the best substitute for a predicate variable, but apparently not intended by Read) is a restriction on first-order logic, as we do not presuppose the availability of such symbols in every formal language.

Read's proposal initiated a small debate. Griffiths (2014), for example, claims that if one pair of introduction and elimination is harmonious, then all other equivalent pairs are so. His claim seems to be based on a "semantic" reading of harmony, exemplified, in particular, in the context of general elimination rules. Using this, he reduces the asserted harmony of Read's suggestion to the standard non-harmonious inference rules. Read (2016) seems to accept Griffiths' claim (not the criticism), which motivates him to provide an improved version of his strong introduction rule (together with an improved elimination rule, which is the congruence rule (I_5) given in two polarities) and to claim that harmony of his improved rules entails the harmony of his first proposal.

Klev (2019) criticises Read's justification of the elimination rules (in both of Read's papers) and uses the opportunity to suggest his own proposal. Motivated by Martin Löf's type theory, he claims that the usual identity rules (the introduction rule in a version proposed by Martin Löf) are harmonious. In order to succeed, he externalises the problematic aspects into a secondary identity relation, the so called definitional identity. Furthermore, the concept of harmony is changed: instead of justifying the elimination rule by the considered introduction rule, the elimination rule is justified in terms of this rule accompanied by finitely many subsequent substitutions (entailed by the secondary identity relation).

We consider neither solution convincing; but a detailed analysis of the debate is beyond the possibilities of these investigations.

Our proposal. Motivated by actual mathematical praxis, Gazzari (2021) proposed the calculus of Natural Calculation (NC), which is an extension of Gentzen's ND by proper term rules permitting a natural representation of calculations.[3] A brief introduction into this calculus is given in section 2. Some of these calculation rules turn out to be introduction and elimination of the equality symbol; additionally, there is a substitution rule available (in the presence of proper relation symbols), which we do not consider as an equality rule, as this symbol does not occur in the formulation of the rule.

In section 3, we argue that the NC versions of the introduction and the elimination of equality are harmonious, as they satisfy an inversion principle as suggested by Prawitz (1971). Therefore, we should consider equality as a logical symbol. Even though the NC substitution rule does not fit into the introduction and elimination schema, we will argue in section 4 that this rule should be considered harmonious insofar this rule is self-inverse.

In section 5, we suggest an extension of the usual BHK interpretation of logic by a new clause for the equality symbol. In this new interpretation of equations, the standard identity rules can be understood as meta-rules expressing the basic properties of calculations (similar to sequent calculus rules expressing properties of derivations). Reflecting the dual character of the equality rules (which are also calculation rules), we suggest additional BHK clauses dealing, in particular, with the elimination of equality (which is the rule permitting calculation steps).

In section 6, we analyse standard ND in the light of the proposed BHK interpretation of the equality symbol. For this purpose, we provide an embedding of NC into standard ND based on the BHK interpretation of equality.

---

[3] It is worth mentioning that Indrzejczak (2021) carried over the idea of proper term rules to Gentzen's sequent calculus; similarly to our discussion of the inversion principle, he investigates the harmony of these new kind of inference rules in the sequent calculus setting.

René Gazzari

In order to represent faithfully the NC version of the introduction of equality, we have to define a new ND inference rule ($I_=$). This rule is admissible in standard ND, but not derivable. The latter can explain the difficulties to find harmonious inference rules for the equality in standard ND. More generally, we will see that none of the standard identity rules of ND (as given in figure 1) should be considered as introduction or as elimination of the equality.

2 The calculus of Natural Calculation

Preceding to our introduction of the calculus of Natural Calculation (NC), we comment briefly on nominal forms and substitution.

Remark 1 (Nominal forms and substitution) *Nominal terms* r (essentially, as introduced by Schütte, 1977) are a generalisation of first-order terms, in which nominal symbols $*_k$ may occur; they are used to mark the position of occurrences of terms in terms and to represent the replacement of these occurrences. $r[t_0, \ldots t_n]$ is the result of a simultaneous replacement of the nominal symbols $*_k$ in r by the kth arguments t_k. r is *unary*, if only $*_0$ occurs in r (at least once, possibly multiple times); consequently, $r[t]$ is a standard term for all standard terms t. r marks the (multiple) position of an occurrence of t in a term r, if $r \equiv r[t]$. The replacement of the marked occurrences by a term s results in $r[s]$.

Analogously, unary *nominal formulae* A represent the position of first-order terms t occurring in a formula $A \equiv A[t]$; as before, $A[s]$ is the result of the replacement of the intended occurrences by a term s. We use the expression *nominal form* to subsume nominal terms and nominal formulae. A more detailed analysis of nominal forms, occurrences and their positions is found in (Gazzari, 2020).

The calculus of Natural Calculation (NC) is the extension of Gentzen's original version of ND (without the standard identity rules given in figure 1), in which a derivation can be generated, additionally, by term assumptions and according to the following term rules.

Definition 1 (Term rules of NC) *The following rules are available in NC:*

$$\dfrac{\begin{array}{c}[t]\\s\end{array}}{t=s}\,(I_=) \quad ; \quad \dfrac{r[s_0] \quad s_0=s_1}{r[s_1]}\,(E_=^+) \quad ; \quad \dfrac{r[s_1] \quad s_0=s_1}{r[s_0]}\,(E_=^-)$$

The elimination rule ($E_=^\pm$) *is given in two polarities. The term premises* $r[s_0]$ *and* $r[s_1]$, *respectively, are the* minor premise *of the respective inference step,* the equation $s_0 = s_1$ is the major premise. Alternatively, the term premises are called the affected terms and the equations the (direct) justifications.

Additionally, the following substitution rule is available for all $n + 1$-ary relation symbols R:

$$\frac{R(t_0, \ldots t_n) \quad \begin{array}{c}[t_0]\\ s_0\end{array} \quad \ldots \quad \begin{array}{c}[t_n]\\ s_n\end{array}}{R(s_0, \ldots s_n)} \ (S)$$

The formula premise is called the affected formula, the calculations from t_k to s_k are the justifications (for all $k \leq n$).

The (complete) discharge of term assumptions is mandatory in all rules permitting such a discharge.

Example 1 (An example derivation) The statement $n + 0 = 0 + n$ (for all natural numbers n) is proved in arithmetics by induction; in the induction step, one has to establish $S(n) + 0 = 0 + S(n)$ under the presupposition of the induction hypothesis. This is done via the following calculation:

$$S(n) + 0 \stackrel{(A_1)}{=} S(n) \stackrel{(A_1)}{=} S(n+0) \stackrel{(IH)}{=} S(0+n) \stackrel{(A_2)}{=} 0 + S(n)$$

The equation $S(n) + 0 = 0 + S(n)$ is obtained by reference to the transitivity of equality. The indications A_1 and A_2 refer to the axioms of addition in Peano arithmetics and IH is the induction hypothesis. The following derivation represents this "informal" calculation in NC:

$$\cfrac{S(x)+0 \quad \cfrac{\cfrac{A_1}{S(x)+0 = S(x)}}{S(x)}(E_=^+) \quad \cfrac{\cfrac{A_1}{x+0=x}}{S(x+0)}(E_=^-) \quad IH}{\cfrac{\cfrac{}{S(0+x)}(E_=^+) \quad \cfrac{A_2}{0+S(x) = S(0+x)}}{0+S(x)}(E_=^-)}$$

The required equations (besides IH) are obtained from the axioms A_1 and A_2 by some elimination of the universal quantifier steps; IH is an assumption, which becomes discharged in the full proof of the proposition.

The equation $S(x) + 0 = 0 + S(x)$ can be derived by an application of $(I_=)$ involving the mandatory discharge of the open term assumption $S(x) + 0$. While the first elimination is applied positively to the full term, the second is applied negatively and affects only a subterm. Analogously, the step using IH affects only a subterm of $S(x+0)$ positively. The last elimination step is negative and affects the full term.

We complement the definition of the term rules by introducing some terminology related to the availability of term rules.

Definition 2 (Terminology) *A derivation* **C** *with term conclusion is called a* calculation. *The open term assumption t of a calculation* **C** *is the* initial term, *the term conclusion s the* final term *and the equation $t = s$ the* result *of* **C**. *Derivations generated by* $(I_=)$ *in the last inference step, thus inferring the result of the respective calculation, are called* closed calculations. *The notion of a path (as given by Prawitz, 1965) is easily generalised.*

Remark 2 (Natural Calculation)
1. NC is equivalent to ND extended by the standard identity rules in intuitionistic and in classical logic.

2. There are some minor differences to the original presentation of NC in (Gazzari, 2021): we omit the redundant *auxiliary calculation (term) rule*. Furthermore, we restrict the substitution rule to relational expressions, but we demand the replacement of all arguments; due to these changes, the substitution rule obtains better proof-theoretic properties.

3. It is worth mentioning that Read's revised version of the elimination of equality rule (Read, 2016) is also given in two polarities. Also, Read's original introduction rule in (Read, 2004) is reminiscent of the NC rule, if we neglect the predicate variable used in his formulation. We comment on Read's revised version of the introduction rule (with two premises) and its (potential) NC analogue when discussing the inversion principle.

Dual perspective on term rules. There are two competing perspectives on the term rules: as reflected by the denominations, the rules $(I_=)$ and $(E_=^\pm)$ can be seen as rules governing the equality symbol. The categorisation of the premises into major and minor depends on this perspective and, therefore, also the notion of paths (as introduced above).

Alternatively, we can see the elimination rule as a rule characterising calculation steps and the introduction rule as an evaluation rule, in which the result of a calculation is inferred. Under this perspective, the affected terms in $(E_=^\pm)$ become the major premises and the direct justifications the minor premises. The resulting concept of a path is called *calculative* path. The *main calculative* path of a calculation starts in the term assumption and ends in the final term of this calculation.

Reconsidering Identity

The equality symbol does not occur in the formulation of the substitution rule; consequently, we do not consider this rule as an equality rule, but only as a calculation rule. Accordingly, the affected formula is considered as the major premise and the final terms of the justifications as minor premises.

The calculative paradigm is relevant in the definition of the basic operations for calculations, which can be explained as operations affecting main calculative paths.

Example 2 (Standard identity rules) The following NC derivations witness the derivability of the standard identity rules (I_1), (I_2), (I_3) and (I_4).

$$\dfrac{[\mathbf{t}]^1}{t=t}\,(I_=\!:\!1) \quad ; \quad \dfrac{\dfrac{[\mathbf{s}]^1 \quad t=s}{\mathbf{t}}\,(E_=^-)}{s=t}\,(I_=\!:\!1)$$

$$\dfrac{\dfrac{[\mathbf{t}]^1 \quad t=s}{\mathbf{s}}\,(E_=^+) \quad s=r}{\dfrac{\mathbf{r}}{t=r}\,(I_=\!:\!1)}\,(E_=^+) \quad ; \quad \dfrac{\dfrac{[\mathbf{r}[t]]^1 \quad t=s}{\mathbf{r}[s]}\,(E_=^+)}{\mathbf{r}[t]=\mathbf{r}[s]}\,(I_=\!:\!1)$$

The proper terms in the derivations (which form the main calculative paths inside the closed calculations) are highlighted for readability reasons.

Definition 3 (Operations on calculations) *The following operations on calculations are easily defined by recursion on the structure of calculations:*

1. *The dual* $\overline{\mathbf{C}}$ *of a calculation* $\mathbf{C}$ *is obtained by inverting the order of the main calculative path of* $\mathbf{C}$ *(bottom up) and swapping the polarities of the involved elimination steps.*

2. *The* m-*variants* $\mathrm{m}[\mathbf{C}]$ *of a calculation* $\mathbf{C}$ *are the result of replacing all terms t in the main calculative path of* $\mathbf{C}$ *by* $\mathrm{m}[t]$ *for all unary nominal terms* m.

Example 3 (Operations on calculations) We provide two examples illustrating the operations on calculations. Let $\mathbf{C}$ be the following calculation:

$$\dfrac{\dfrac{\dfrac{a \quad a=b}{b}\,(E_=^+) \quad c=b}{c}\,(E_=^-) \quad a=c}{a}\,(E_=^-)$$

We apply both operations on $\mathbf{C}$:

1. *Dualisation:* $\overline{\mathbf{C}}$ *is given as follows:*

$$\dfrac{a \quad \dfrac{a=c}{c}\,(E_{=}^{+})\quad c=b}{\dfrac{b}{a}\,(E_{=}^{-})}\,(E_{=}^{+}) \quad a=b$$

2. m-Variant: $\mathtt{m}[\mathbf{C}]$ with $\mathtt{m} \equiv *_0 + a$ is given as follows:

$$\dfrac{\dfrac{\mathtt{m}[a] \equiv a+a \quad a=b}{\mathtt{m}[b] \equiv b+a}\,(E_{=}^{+}) \quad c=b}{\dfrac{\mathtt{m}[c] \equiv c+a}{\mathtt{m}[a] \equiv a+a}\,(E_{=}^{+})}\,(E_{=}^{-}) \quad a=c$$

Observation 1 (Operations on calculations) *The following observations are easily proved by induction on the structure of calculations: If $t = s$ is the result of $\mathbf{C}$, then $s = t$ is the result of $\overline{\mathbf{C}}$ and $\mathtt{m}[t] = \mathtt{m}[s]$ is the result of $\mathtt{m}[\mathbf{C}]$ (both results depending on the unchanged justifications of $\mathbf{C}$).*

3 Satisfaction of the inversion principle

We base our investigations on an inversion principle as formulated by Prawitz (1971, p. 246f, quotation marks by Prawitz):

> In other words, a proof of the conclusion of an elimination is already "contained" in the proofs of the premisses when the major premiss is inferred by introduction.

This version of the inversion principle is essentially a reformulation of the principle already stated in (Prawitz, 1965), but avoiding the technicalities. Even though the inversion principle is formulated with respect to formula rules, we do not see any obstacle to carry over this principle to term rules.

The concept of containment (set in quotation marks) is slightly vague and open for interpretations.[4] Our strategy to deal with this vagueness is based on the idea that the usual inference rules of the standard logical operators (in particular, implication and universal quantifier) satisfy the inversion principle: in order to justify containment in the case of term rules, it is sufficient to argue that accepting this containment is comparable to accepting containment in the case of the standard operators.

---

[4] From the perspective of our reviewer, the subsequent discussion illustrates how problematic the concept of "containment" is, if it is taken too literally.

Reconsidering Identity

Analysis of the equality rules. The elimination rule ($E_{=}^{\pm}$) for the equality is given in two polarities as follows:

$$\frac{\mathbf{r}[s_0] \quad s_0 = s_1}{\mathbf{r}[s_1]} \; (E_{=}^{+}) \quad ; \quad \frac{\mathbf{r}[s_1] \quad s_0 = s_1}{\mathbf{r}[s_0]} \; (E_{=}^{-})$$

The major premise of the elimination rule (in both polarities) is the equation $s_0 = s_1$. A canonical proof of the major premise is obtained from a calculation **C** as follows:

$$\mathbf{D} \equiv \begin{array}{c} [s_0]^1 \\ \mathbf{C} \\ \dfrac{s_1}{s_0 = s_1} \; (I_0{:}1) \end{array}$$

The conclusions $\mathbf{r}[s_1]$ and $\mathbf{r}[s_0]$ of the applications of the elimination rules, respectively, can be obtained from the canonical proof **D** of the major premise $s_0 = s_1$ as follows:

$$\begin{array}{c} \mathbf{r}[s_0] \\ \mathbf{r}(\mathbf{C}) \\ \mathbf{r}[s_1] \end{array} \quad ; \quad \begin{array}{c} \mathbf{r}[s_1] \\ \mathbf{r}(\overline{\mathbf{C}}) \\ \mathbf{r}[s_0] \end{array}$$

Recall that $\mathbf{r}(\mathbf{C})$ is the **r**-variant of **C**, where all the terms t in the main calculative path are replaced by their **r**-variants $\mathbf{r}[t]$, and that $\overline{\mathbf{C}}$ is the dual of **C**, which is the calculation **C** read bottom up.

In order to satisfy the inversion principle, the calculations $\mathbf{r}(\mathbf{C})$ and $\mathbf{r}(\overline{\mathbf{C}})$, respectively, have to be "contained" in the proof **D**. Subsequently, we consider this "containment" in dependence on the non-logical symbols available in the underlying formal language $\mathfrak{L}$.

Pure languages. We presuppose that $\mathfrak{L}$ is the pure first-order language without any non-logical symbols.

Consequently, first-order variables x are the only terms of $\mathfrak{L}$. In particular, the affected terms $\mathbf{r}[s_0]$ and $\mathbf{r}[s_1]$ are variables x. This means that the unary nominal term $\mathbf{r}$ must be the nominal symbol $*_0$ (this is $\mathbf{r} \equiv *_0$) and, therefore, $\mathbf{r}[t] \equiv t$ for all terms t of $\mathfrak{L}$. As a consequence, we obtain $\mathbf{r}(\mathbf{C}) \equiv \mathbf{C}$.

In the positive case, the inversion principle is perfectly satisfied, as $\mathbf{r}(\mathbf{C}) \equiv \mathbf{C}$ is contained in **D** exactly as it is. In the negative case, the situation is slightly more involved: the calculation $\mathbf{r}(\overline{\mathbf{C}}) \equiv \overline{\mathbf{C}}$ is obtained from **C** by inverting the main calculative path bottom up and swapping, additionally, the polarity of the applied calculation rules (but using the same justifications). As $\overline{\mathbf{C}}$ is a mere rearrangement of $\mathbf{C}$, it seems reasonable to accept $\overline{\mathbf{C}}$ as "contained" in $\mathbf{D}$. The latter means that the inversion principle is also satisfied in the negative case.

The price we pay is to accept that the rearrangement of $\mathbf{C}$ (as prescribed by the dualisation function) is not affecting containment. The situation is similar to the conversion needed for the implication, where we have to compose two derivations, which is a simple kind of rearrangement.

Languages with constant symbols. If we permit constant symbols c in the underlying language $\mathfrak{L}$, but no proper function symbols f (with arity greater than zero), then we can conclude, analogously to the case above, that $\mathbf{r} \equiv *_0$. The rest of the analysis remains unchanged.

Languages with proper function symbols. We presuppose that proper function symbols f (with arity greater than zero) are available in the underlying formal language $\mathfrak{L}$.

Due to the availability of function symbols, the situation changes: we cannot conclude $\mathbf{r} \equiv *_0$. This means that $\mathbf{r}(\mathbf{C})$ results from replacing the terms t in the main calculative path of $\mathbf{C}$ by the terms $\mathbf{r}[t]$ (which are, in the general case, different from t); the justifications of the calculation steps remain unchanged. As before, it seems reasonable to accept $\mathbf{r}(\mathbf{C})$ as "contained" in $\mathbf{C}$ (and, therefore, in $\mathbf{D}$) and to claim that the inversion principle is satisfied: in the case of the universal quantifier, we have to accept that the eigenvariable is replaced at all of its occurrences in the canonical proof. The latter means that a substitution of terms does not contradict "containment".

It seems that the rules for the universal quantifier behave even worse than the calculation rules: perfect harmony is already lost in the presence of constant symbols and not only in the presence of complex terms; moreover, the term t by which the eigenvariable has to be replaced is only determined by the application of the elimination rule, whereas the context $\mathbf{r}$ in which we have to embed the calculation $\mathbf{C}$ is already present in the premises of the elimination of the equality rule.

This means that in the positive case, the proof $\mathbf{r}[\mathbf{C}]$ of the major premise is, indeed, contained in the canonical proof $\mathbf{D}$ and, therefore, that the inversion principle is satisfied. In the negative case, we only have to accept additionally (as before) that dualisation does not change the situation.

Languages with relation symbols. The presence of relation symbols does not affect our analysis of the equality rules.

Harmony of the equality rules. We summarise: accepting some variations of a strict containment, similar to those usually accepted, the elimination of equality rule ($E_=^\pm$) satisfies the Prawitz inversion principle with respect to the introduction of equality rule ($I_=$). This means that the rules governing the equality symbol are harmonious and that we should consider this symbol as a logical symbol.

Remark 3 (Alternative introduction rule)   As mentioned, we find in (Read, 2016) an alternative introduction rule for the equality. The NC version of this improved rule would look like:

$$\dfrac{\overset{[t]}{s} \quad \overset{[s]}{t}}{t = s} \, (I_=^*)$$

Clearly, due to the existence of dual calculations, the rules ($I_=$) and ($I_=^*$) permit to infer the same equations.

Regarding the inversion principle, we have to admit that ($I_=^*$) seems to be superior. We do not have to accept, in the negative case, that $\overline{\mathbf{C}}$ is "contained" in a canonical proof $\mathbf{D}$ (via a rearrangement of $\mathbf{C}$), but we have a calculation $\mathbf{C'}$ from s to t explicitly contained in $\mathbf{D}$ (which can even be a calculation different from $\overline{\mathbf{C}}$). This means that the commitment for accepting containment becomes smaller.

Besides this observation, we still opt for the simple introduction rule ($I_=$), as this version of the rule is natural (with respect to mathematical praxis, when dealing only with equalities). The ($I_=^*$) becomes relevant, as soon as we consider smaller-than calculations (characterised by different elimination rules). In this setting, the introduction rule ($I_\leq$) for smaller-than statements $t \leq s$ would correspond to ($I_=$), while ($I_=^*$) would be required for inferring $t = s$. A detailed analysis of this setting with two (dependent) primitive relation symbols has to be left to future work.

4   Analysing the substitution rule

As mentioned in the introduction of NC (section 2), we do not consider the substitution rule as an introduction or an elimination rule. Consequently, there is no point in asking whether the rule is in harmony with another rule.

Furthermore, we do not consider the substitution rule as an equality rule (as no equality symbol occurs in the formulation of this rule), but only as a calculation rule. But nevertheless, there are aspects regarding this rule worth mentioning in our analysis of harmony.

Reducibility of the substitution rule. As the composition $\mathbf{C} \circ \mathbf{C}'$ of two calculations $\mathbf{C}$ and $\mathbf{C}'$ is again a calculation (provided that the final term of $\mathbf{C}$ and the initial term of $\mathbf{C}'$ are equal), we can reduce two subsequent applications of the substitution rule into a single application. In the case of a unary relation symbol R:

$$\dfrac{\dfrac{R(t) \quad \begin{matrix}[t]\\ \mathbf{C}_t \\ s\end{matrix}}{R(s)} \quad \begin{matrix}[s]\\ \mathbf{C}_s \\ r\end{matrix}}{R(r)} \quad \rightsquigarrow \quad \dfrac{R(t) \quad \begin{matrix}[t]\\ \mathbf{C}_t \circ \mathbf{C}_s \\ r\end{matrix}}{R(r)}$$

Observe that we have to compose the justifications, which is a rearrangement of derivations similar to that needed with respect to the implication.

Furthermore, we can eliminate a single application of the substitution rule, if all justifying calculations $\mathbf{C}_k$ satisfy the condition that their initial terms t_k are syntactically equal to their final terms s_k. In the case of a unary relation symbol R:

$$\dfrac{R(t) \quad \begin{matrix}[t]\\ \mathbf{C} \\ t\end{matrix}}{R(t)} \quad \rightsquigarrow \quad R(t)$$

Self-inverse substitution rule. Combining both observations, we may consider the substitution rule to be self-inverse in the following sense: whenever we apply the substitution rule (justified by calculations $\mathbf{C}_k$), we can infer the affected formula $R(\vec{t})$ via a second application of the substitution rule (using the dual calculations $\overline{\mathbf{C}_k}$ as justifications). Finally, both applications can be eliminated. In the case of unary relation symbols R:

$$\dfrac{R(t) \quad \begin{matrix}[t]\\ \mathbf{C} \\ s\end{matrix}}{R(s)} \quad \rightsquigarrow \quad \dfrac{\dfrac{R(t) \quad \begin{matrix}[t]\\ \mathbf{C} \\ s\end{matrix}}{R(s)} \quad \begin{matrix}[s]\\ \overline{\mathbf{C}} \\ t\end{matrix}}{R(t)} \quad \rightsquigarrow \quad \dfrac{R(t) \quad \begin{matrix}[t]\\ \mathbf{C} \circ \overline{\mathbf{C}} \\ t\end{matrix}}{R(t)} \quad \rightsquigarrow \quad R(t)$$

We emphasise that this notion of being self-inverse is based on an equality of derivations finer than the usual identity stipulated by normalisation, as it is term sensitive. Furthermore, this property can be seen as a kind of harmony (independent of the introduction-elimination dichotomy).

Remark 4 (Versions of the substitution rule)   The reducibility of subsequent applications of the substitution rule is the reason to opt for this version of the rule and it depends strongly on its formulation:

- Not demanding the replacement of full arguments t_k (but permitting the replacement of subterms), subsequent justifications would not be composable (as final terms and initial terms could be different).

- Permitting the replacement of only a single argument, we would lose again composability of the justifications (for the same reason).

- Accepting equations (instead of calculations) as justification, we would lose composability: we would have to apply explicitly transitivity (in some derivation steps) to obtain the suitable justifying equation for the contracted application.

In particular, rule (I_5) is not reducible and, therefore, not self-inverse.

5   Constructive interpretation of the equality

According to Gentzen (1934), the meaning of the logical operators is given by their introduction rules. This idea is substantiated by the BHK interpretation of logic, which is an (informal) interpretation of the logical operators for intutitionistic or, more generally, constructive logic; it provides the grounds on which a (complex) formula can be asserted. Our analysis of the equality rules motivates a BHK interpretation of the equality symbol (of equations) based on the first-order terms constituting the respective equation:

- A proof of an equation $t = s$ is a closed (equality) calculation from the term t to the term s.

It is immediate that the introduction of equality rule $(I_=)$ of NC exemplifies the suggested reading of identity. In contrast to the BHK interpretation of connectives and quantifier, the meaning of an equation $t = s$ is not explained in terms of *proofs* of its constituents t and s (as first-order terms cannot have proofs). But still, the explanation is given in terms of its constituents using calculations, which can be seen as the subatomic counterparts of proofs.[5]

Reformulation (BHK interpretation). Slightly reformulating the BHK clause for the equality, we obtain the following clause for equations:

- An equation $t = s$ is the statement asserting the existence of an equality calculation from the term t to the term s. $(\star)$

This clause will be used in our analysis of ND from the perspective of NC in the next section.

Remark 5 (Free variables)   It is tempting to claim that in cases, where t or s contain free variables, the reformulation is, in general, not correct. Considering, for example, the equation $x + 0 = 0 + x$ (†) from example 1, one would argue that there is a calculation from $t + 0$ to $0 + t$ for every constant term t (of arithmetics), but not from $x + 0$ to $0 + x$, as (†) has to be proved by induction. Consequently, an equation expresses more than the existence of a calculation.

We do not agree with such an argumentation. First, we have to mention here that there is only in the trivial case, where t is 0, a calculation from $t + 0$ to $0 + t$ in the strict sense that the calculation does not depend on assumptions.[6] A calculation, for example, from $1 + 0$ to $0 + 1$ requires derived equations as justifications, in this example derived from the axioms of PA. The same is true for a calculation from $x + 0$ to $0 + x$, with the only difference that the justifying equation is derived, in particular, with the help of an instance of the induction schema. The latter is not needed, if t is a constant term.

---

[5]The (traditional) BHK interpretation of proofs as functions can be made precise via the Curry-Howard correspondence into simply typed λ-calculus. A quite natural (and relevant) question arising here is, in which way the concept of a calculation can be made precise. A preliminary answer to this question is that a simply typed λ-calculus, corresponding in the Curry-Howard sense to NC, is under development. In this calculus, there are two kinds of terms: besides the (usual) proof terms (typed by formulae), there are also calculation terms (built from a special variable representing the term assumptions and typed by first-order terms).

However, this answer seems as useful as saying that calculations are derivations with term conclusion while proofs are derivations with formula conclusion. A more profound analysis of the concept of calculations is necessary, but beyond the possibilities of these investigations and left to future work.

[6]This is the case, as an equation $t = s$ is provable, if and only if t and s are syntactically equal. A proof of this result is beyond the possibilities of these investigations.

Reconsidering Identity

One reason for doubting the existence of a calculation from $x + 0$ to $0 + x$ is that the (paradigmatic) calculation is quite trivial, namely a single calculation step with a huge proof (in PA) of the justifying equation:

$$\cfrac{x+0 \qquad \cfrac{\ldots \to \forall x.x+0 = 0+x \qquad \cfrac{\text{PA}}{A(0) \land \forall x.(A(x) \to A(S(x)))}\ (E_\to)}{\cfrac{\forall x.x+0 = 0+x}{x+0 = 0+x}\ (E_\forall)}}{0+x}\ (E_=^\pm)$$

Here, $A(x) \equiv x + 0 = 0 + x$ and the formula notated with "..." is the instance of the induction schema with respect to A (and included in PA). A less trivial calculation involving variables is the calculation from $S(x) + 0$ to $0 + S(x)$ given in example 1, not using the induction schema, but (†) as induction hypothesis. Also, the calculations (within the formula derivations) of example 2 are based on variables.

Another source for doubts could be the fact that we derive first the justifying equation $x + 0 = 0 + x$ allowing for the trivial calculation discussed above (and being, additionally, the result of this trivial calculation). And this equation is, indeed, not obtained in a canonical proof ending with an introduction of the equality, but by an induction, which means as the conclusion inside of an elimination part of the derivation. But this observation is not specific for equations and calculations, but is a general phenomenon for hypothetical reasoning, which is, in particular, reasoning inside of formal theories as PA.[7]

Meaning of equality. In contrast to the expectations, the BHK clause for the equality symbol (the respective introduction rule) seems not to determine the (full) meaning of the characterised symbol. Assuming that the canonical properties of identity constitute the meaning of the equality symbol, we observe:

1. Reflexivity (given by the possibility of immediate introduction) and transitivity (given by the possibility of extending calculations) depend on the introduction of equality rule $(I_=)$.

2. Symmetry depends on the elimination rule $(E_=^\pm)$; more precisely, on the fact that positive and negative applications can affect the *same*

---

[7]Analogously, we would claim that the implication $A \to \bot$ is the statement asserting the existence of a proof from A to the absurdity.

occurrences of subterms in the affected terms. Congruence with respect to function symbols depends on the fact that we may affect in a calculation step *all* subterms of the affected term. (This would be different with anti-symmetric smaller-than calculations.)

3. Finally, we have to mention that congruence with respect to relation symbols and arbitrary formulae (in the presence of relation symbols different from the equality) depends on the substitution rule.

Complementing the BHK interpretation. It seems that the BHK interpretation, as given so far, is incomplete; not only because of our observations regarding the meaning of equality, but also as calculations are a new term (not dealt with in the usual BHK clauses) and as the substitution rule does not fit into the schema. The situation could be improved by extending the BHK interpretation by the following non-standard clauses:

1. An equality calculation step from a term $r[t]$ to a term $r[s]$ is the justified replacement of t by s; the equations $t = s$ and $s = t$ can serve as justification.

2. Assuming a term yields a calculation as well as extending a calculation by an (additional) calculation step.

3. Equality calculations from t_k to s_k justify the replacement of the arguments $\vec{t}$ in $R(\vec{t})$ by $\vec{s}$.

It is worth mentioning that we do not consider clause (1) as a BHK clause for the equality symbol (dealing with its elimination), but as a clause for the calculation step rule (which coincides with the elimination, but belongs to the alternative paradigm of calculations). In particular, we do not see any reason to demand that the usual BHK clauses for connectives and quantifier become complemented by an additional clause for the elimination rules.

Without analysing Brouwer's exact conception of construction, it seems that our suggestions are in line with van Atten (2022), when he demands for BHK clauses of atomic formulae:

> A proof of an atomic proposition A is given by presenting a mathematical construction in Brouwer's sense that makes A true.

6 Revisiting Natural Deduction

We provide an interpretation of the standard identity rules of ND in terms of the reformulation ($\star$) of the BHK interpretation of the equality symbol. This interpretation is used as a guideline for an embedding of NC into ND, which permits an analysis of the standard identity rules of ND from the perspective of NC.

Interpretation of the standard identity rules. The interpretation ($\star$) of equations asserting the existence of calculations yields an interpretation of the standard ND identity rules as meta-rules expressing essential properties of calculations.

- (I_1) - *reflexivity:* Every trivial calculation t (which is the term assumption t) can be evaluated and transformed this way into a proof of the trivial equation $t = t$.

- (I_2) - *symmetry:* Every calculation **C** from t to s can be dualised.

- (I_3) - *transitivity:* Two calculations **C** from t to s and **C'** from s to r can be composed to a calculation **C''** from t to r.

- (I_4) - *congruence (term):* Every calculation **C** from t to s can be embedded into a broader context m, which yields a calculation m[**C**] from m[t] to m[s].

As we do not consider the substitution rule as an equality rule, we abstain here from suggesting an interpretation of (I_5).

Embedding NC into ND. We use the idea of representing NC calculations **C** from t to s in ND by equations $t = s$ (according to the BHK clause for equations). More schematically, the embedding $\varepsilon : \text{NC} \hookrightarrow \text{ND}$ is given according to the following mapping:

$$\varepsilon : \begin{array}{c} t \\ \mathbf{C} \\ s \end{array} \mapsto \begin{array}{c} t = t \\ \mathbf{D} \\ t = s \end{array}$$

Subsequently, we discuss the details.

Representation (term assumption). Term assumptions t of NC are represented by the formula assumption $t = t$.

Representation (elimination of equality). The NC elimination of the equality rule $(E^\pm_=)$ is represented by the following formula rule $(\mathbf{E}^\pm_=)$ (given in two polarities) in ND:

$$\dfrac{t = \mathbf{r}[s_0] \quad s_0 = s_1}{t = \mathbf{r}[s_1]}\ (\mathbf{E}^+_=) \quad ; \quad \dfrac{t = \mathbf{r}[s_1] \quad s_0 = s_1}{t = \mathbf{r}[s_0]}\ (\mathbf{E}^-_=)$$

$(\mathbf{E}^+_=)$ is a special case of the rule (I_5); $(\mathbf{E}^-_=)$ can be derived via an additional application of symmetry. But, more interestingly, both polarities of $(\mathbf{E}^\pm_=)$ are also derivable in standard ND without an application of (I_5):

$$\dfrac{t = \mathbf{r}[s_0] \quad \dfrac{s_0 = s_1}{\mathbf{r}[s_0] = \mathbf{r}[s_1]}\ (I_4)}{t = \mathbf{r}[s_1]}\ (I_3) \quad ; \quad \dfrac{t = \mathbf{r}[s_1] \quad \dfrac{\dfrac{s_0 = s_1}{s_1 = s_0}\ (I_2)}{\mathbf{r}[s_1] = \mathbf{r}[s_0]}\ (I_4)}{t = \mathbf{r}[s_0]}\ (I_3)$$

In the negative case, we may alternate the applications of (I_2) and (I_4). It is worth mentioning that the derivability proofs can be read as instructions for normalising NC proofs (by applying the $(\star)$ interpretation of equations).

Representation (introduction of equality). The NC introduction of equality rule $(I_=)$ is represented by the following non-standard formula rule $(\mathbf{I}_=)$:

$$\dfrac{[t = t] \atop t = s}{t = s}\ (\mathbf{I}_=)$$

The discharge of the formula assumption $t = t$ is mandatory and restricted to that single occurrence of the formula assumption $t = t$, which corresponds to the term assumption t. The reflexivity rule (I_1) is a special case of $(\mathbf{I}_=)$.

Obviously, $(\mathbf{I}_=)$ is not derivable from the standard identity rules in ND, but admissible: whenever we have to discharge an assumption $t = t$, we may convert this derivation into a derivation, in which the respective assumption is replaced by an application of (I_1).

Analysis (non-standard introduction). The effect of the non-standard introduction rule $(\mathbf{I}_=)$ is to represent in ND the evaluation of a calculation (which is the introduction of an equality) by discharging a formula assumption $t = t$. If we want to avoid the application of this non-standard rule, the discharge has to be permitted by a different rule. The only possible candidate is the reflexivity rule (I_1) (applied immediately at the begin of the ND representation of a calculation). This means that the introduction of equality rule $(I_=)$ of NC would be represented in ND even before it is applied.

In other words: ND does not provide canonical means to distinguish between ongoing NC calculations and evaluated calculations; there is no canonical representative of the introduction of the equality rule $(I_=)$. Of course, we could represent the introduction of the equality rule $(I_=)$, for example, by a double application of the symmetry rule (I_2); but such a solution is artificial and does not represent faithfully NC.

Representation (substitution rule). The representation of the substitution rule (S) in ND is straightforward by the following formula rule (**S**):

$$\frac{R(\vec{t}) \quad \begin{array}{c}[t_0 = t_0]\\ t_0 = s_0\end{array} \quad \ldots \quad \begin{array}{c}[t_n = t_n]\\ t_n = s_n\end{array}}{R(\vec{s})} \text{ (S)}$$

There are obvious differences to the congruence (formula) rule (I_5): (**S**) affects only atomic relational expressions (instead of arbitrary formulae) and requires the replacement of all arguments (instead of multiple occurrences of a single term). But these differences reflect only our conceptual choices when designing the calculus; we could have defined analogous NC rules. The crucial difference to (I_5) is that (**S**) deals with ongoing calculations, which become closed by an application, while (I_5) is indifferent in this regard.

Analysis (standard identity rules). We summarise our observations:

1. As the NC introduction of equality is not representable in standard ND, none of the standard identity rules should count as the introduction of equality.

2. The rule (I_1) is a strictly weaker special case of the introduction rule.

3. The congruence (formula) rule (I_5) operates without distinguishing ongoing calculations from closed ones and subsumes, partially, the elimination rule and the ND version of the substitution rule. Therefore, (I_5) should not be considered as a pure elimination rule. Furthermore, as the elimination rule is derivable in the fragment of ND without (I_5), (I_5) should not be considered at all as the elimination of equality.

4. The representation of the NC elimination of equality rule requires (I_2), (I_3) and (I_4) altogether. Consequently, none of these standard rules should count as the elimination of equality.

7 Conclusion

We conclude this paper with some remarks regarding the related work considered in the introduction.

Our analysis contradicts Griffiths' claim that equivalent sets of rules are co-harmonious, as we clearly assert that the NC rules are harmonious, while this is not the case for the standard identity rules. We assume that this discrepancy can be explained in terms of the presupposed type of inversion principle. Griffiths considers a "semantic" harmony; his argumentation is based on the formulae, which can be derived in one way or another. The concept of harmony used here depends only on the syntactic shape of the involved derivations; a canonical derivation must consist of sufficient syntactic material to generate a proof of the conclusion of a derivation ending with an elimination step. A detailed analysis of these differences could yield a better understanding of the relationship between the different kinds of inversion principle.

The parallelism between Read's second-order based rules and the NC rules is already considered. We would like to add the observation that Read's rule are the result of an incomplete decomposition of Leibniz's identity into inference rules. As NC was not available, he could only decompose the standard logical operators, but not equations. The latter becomes possible with the help of the proper term rules of NC and yields a calculative version of the congruence (formula) rule (I_5) and a second-order introduction of equality rule similar to its NC version.

There is also a parallelism to Klev, insofar a substitution rule plays a central role in both accounts, even though their use is contrary. While Klev is using the substitution rule to integrate his secondary identity relation, the substitution rule of NC is based on exactly the same calculations, which establish the introduction of equations. Furthermore, by modifying the notion of canonical proofs, the substitutions become an extension of the introduction rule in Klev's account. In NC, the substitution rule is clearly separated from the introduction rule and only required in the presence of proper non-logical relation symbols.

References

Gazzari, R. (2020). *Formal Theories of Occurrences and Substitutions* (Unpublished doctoral dissertation). University of Tübingen.

Gazzari, R. (2021). The Calculus of Natural Calculation. *Studia Logica*, *109*, 1375–1411.

Gentzen, G. (1934). Untersuchungen über das logische Schließen. *Mathematische Zeitschrift*, *39*, 176-210.

Gentzen, G. (1969). Investigations into logical deduction. In M. E. Szabo (Ed.), *The Collected Papers of Gerhard Gentzen* (p. 68-131). North-Holland, Amsterdam.

Griffiths, O. (2014, September). Harmonious rules for identity. *The Review of Symbolic Logic*, *7*(3), 499 - 510.

Indrzejczak, A. (2021). A Novel Approach to Equality. *Synthese*, *199*(1), 4749–4774.

Klev, A. (2019). The Harmony of Identity. *Journal of Philosophical Logic*, *48*, 867-884.

Prawitz, D. (1965). *Natural Deduction: A Proof-Theoretical Study*. Almqvist & Wiksell, Stockholm.

Prawitz, D. (1971). Ideas and results in proof theory. In J. E. Fenstad (Ed.), *Proceedings of the Second Scandinavian Logic Symposium (Oslo 1970)* (p. 235-308). North-Holland, Amsterdam.

Read, S. (2004). Identity and Harmony. *Analysis*, *64*(2).

Read, S. (2016, June). Harmonic inferentialism and the logic of identity. *The Review of Symbolic Logic*, *9*(2), 408 - 420.

Schütte, K. (1977). *Proof Theory* (Vol. 225). Springer-Verlag Berlin Heidelberg.

van Atten, M. (2022). The Development of Intuitionistic Logic. In E. N. Zalta (Ed.), *The Stanford Encyclopedia of Philosophy* (Summer 2022 ed.). Metaphysics Research Lab, Stanford University.

René Gazzari
CMAT, University of Minho
Portugal
E-mail: `elbron@gmx.net`

Preconditionals

WESLEY H. HOLLIDAY[1]

Abstract: In recent work, we introduced a new semantics for conditionals, covering a large class of what we call *preconditionals*. In this paper, we undertake an axiomatic study of preconditionals and subclasses of preconditionals. We then prove that any bounded lattice equipped with a preconditional can be represented by a relational structure, suitably topologized, yielding a single relational semantics for conditional logics normally treated by different semantics, as well as generalizing beyond those semantics.

Keywords: conditionals, Heyting algebras, ortholattices, orthomodular lattices, Sasaki hook, indicatives, counterfactuals, flattening, relational frames

1 Introduction

Conditionals in their different flavors—material, strict, indicative, counterfactual, probabilistic, constructive, quantum, etc.—have long been of central interest in philosophical logic (see Egré and Rott 2021 and references therein). In this paper, based on a talk at Logica 2023, we further investigate a new approach to conditionals introduced in our recent work on the representation of lattices with conditional operations (Holliday 2023, § 6).

We define a *preconditional* $\to$ on a bounded lattice to be a binary operation satisfying five natural axioms, each of which we show to be independent of the others (Section 2.1). We also consider the properties of the associated negation defined by $\neg a := a \to 0$ (Section 2.2). Familiar examples of bounded lattices equipped with a preconditional include Heyting algebras (Section 2.3), ortholattices with the Sasaki hook (Section 2.4), and Lewis-Stalnaker-style conditional algebras satisfying the so-called flattening axiom (Section 2.5). We characterize these classes axiomatically in terms of additional independent axioms they satisfy beyond those of preconditionals.

We then show (Section 3) that every bounded lattice equipped with a preconditional can be represented using a relational structure $(X, \lhd)$, suitably topologized. This yields a single relational semantics for conditional logics

---

[1]Thanks to Yifeng Ding, Matt Mandelkern, Guillaume Massas, Alex Rathke, and Snow Zhang for helpful comments.

normally treated by different semantics, as well as a generalization beyond those semantics. We conclude (Section 4) with some suggested directions for further development of this approach to conditionals.

2 Preconditionals

2.1 The axioms and their independence

The definition of a preconditional from Holliday 2023 was discovered through an attempt to axiomatize the class of lattices with an implication operation amenable to a relational representation described in Section 3. However, here we will begin with axiomatics and turn to representation only at the end.

Definition 1 *Given a bounded lattice L, a* preconditional *on L is a binary operation $\to$ on L satisfying the following for all $a, b, c \in L$:*

1. $1 \to a \leq a$;
2. $a \wedge b \leq a \to b$;
3. $a \to b \leq a \to (a \wedge b)$;
4. $a \to (b \wedge c) \leq a \to b$;
5. $a \to ((a \wedge b) \to c) \leq (a \wedge b) \to c$.

Arguably all of the axioms are intuitively valid for both indicative conditionals and counterfactual conditionals in natural language, but we will not make that case here. Instead, let us begin with the following easy check.

Fact 1 *The axioms of preconditionals are mutually independent.*

Proof. For each axiom, we provide a lattice with a binary operation $\to$ in which the axiom does not hold but it is easy to check that the other axioms do.

For axiom 1, consider the two-element lattice on $\{0, 1\}$ with $\to$ defined by $a \to b = 1$. Since $1 \to 0 = 1 \not\leq 0$, axiom 1 does not hold.

For axiom 2, consider the lattice on $\{0, 1\}$ with $\to$ defined by $a \to b = 0$. Since $1 \wedge 1 = 1 \not\leq 0 = 1 \to 1$, axiom 2 does not hold.

For axiom 3, consider the lattice on $\{0, 1\}$ with $\to$ defined by $a \to b = b$. Since $0 \to 1 = 1 \not\leq 0 = 0 \to 0 = 0 \to (0 \wedge 1)$, axiom 3 does not hold.

For axiom 4, consider the lattice with $\to$ on the left of Figure 1. Since $0 \to (1 \wedge 0) = 0 \to 0 = 1 \not\leq 1/2 = 0 \to 1$, axiom 4 does not hold.

Finally, for axiom 5, consider the lattice with $\to$ on the right of Figure 1. Since $1/2 \to ((1/2 \wedge 0) \to 0) = 1/2 \to (0 \to 0) = 1/2 \to 0 = 1 \not\leq 0 = 0 \to 0 = (1/2 \wedge 0) \to 0$, axiom 5 does not hold. □

Preconditionals

	→	0	1/2	1
0		1	1/2	1/2
1/2		0	1/2	1/2
1		0	1/2	1

	→	0	1/2	1
0		0	0	0
1/2		1	1	1
1		0	1/2	1

Lattice L: $0 < 1/2 < 1$.

Figure 1: Left and center: lattice L and $\to$ demonstrating independence of axiom 4. Right: $\to$ demonstrating independence of axiom 5.

2.2 Precomplementation

The preconditional axioms settle some basic properties of the negation operation defined from $\to$ by $\neg x := x \to 0$.

Proposition 1 *Let L be a bounded lattice with a preconditional $\to$. Then defining $\neg x := x \to 0$, we have:*

1. $a \leq b$ *implies* $\neg b \leq \neg a$;

2. $\neg 1 = 0$.

Proof. For part 1, if $a \leq b$, then we have

$$\begin{aligned} b \to 0 &\leq b \to ((b \wedge a) \to 0) &&\text{by axiom 4 of preconditionals} \\ &\leq (b \wedge a) \to 0 &&\text{by axiom 5 of preconditionals} \\ &\leq a \to 0 &&\text{since } b \wedge a = a \text{ from } a \leq b. \end{aligned}$$

Part 2 is immediate from axiom 1 of preconditionals. □

Following Holliday 2023, we call a unary operation $\neg$ satisfying parts 1 and 2 of Proposition 1 a *precomplementation*. Given a precomplementation, we can induce a preconditional as follows—an idea to which we will return in the context of ortholattices in Section 2.4.

Proposition 2 *Let L be a bounded lattice equipped with a precomplementation $\neg$. Then the binary operation $\to$ defined by*

$$a \to b := \neg a \vee (a \wedge b)$$

is a preconditional.

Proof. Using the definition of $\to$, the axioms of preconditionals become:

1. $\neg 1 \vee (1 \wedge a) \leq a$;
2. $a \wedge b \leq \neg a \vee (a \wedge b)$;
3. $\neg a \vee (a \wedge b) \leq \neg a \vee (a \wedge (a \wedge b))$;
4. $\neg a \vee (a \wedge (b \wedge c)) \leq \neg a \vee (a \wedge b)$;
5. $\neg a \vee (a \wedge (\neg(a \wedge b) \vee ((a \wedge b) \wedge c))) \leq \neg(a \wedge b) \vee ((a \wedge b) \wedge c)$.

Axiom 1 holds given the property of precomplementations that $\neg 1 = 0$. Axioms 2-4 follow from the axioms for lattices. Axiom 5 holds given the property of precomplementations that $a \wedge b \leq a$ implies $\neg a \leq \neg(a \wedge b)$. □

2.3 Heyting implication

As suggested in Section 1, several familiar conditional operations are examples of preconditionals. Our first example is the Heyting implication in Heyting algebras. Consider the following axioms:

- modus ponens (MP): $a \wedge (a \to b) \leq b$;
- weak monotonicity: $b \leq a \to b$.

Fact 2

1. *Modus ponens is independent of the axioms of preconditionals plus weak monotonicity.*

2. *Weak monotonicity is independent of the axioms of preconditionals plus modus ponens.*

Proof. For modus ponens, consider the lattice with $\to$ in Figure 2. We have $b \wedge (b \to a) = b \wedge 1 = b \not\leq a$, so modus ponens does not hold, but one can check that the other axioms do.

For weak monotonicity, consider the two-element lattice on $\{0, 1\}$ with $\to$ defined by $a \to b = a \wedge b$. Since $1 \not\leq 0 = 0 \to 1$, weak monotonicity does not hold, but one can check that the other axioms do. □

We can characterize Heyting implications as preconditionals satisfying the two axioms above.

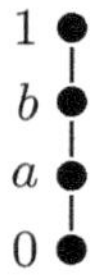

Preconditionals

```
1 ●
b ●
a ●
0 ●
```

→	0	a	b	1
0	1	1	1	1
a	0	1	1	1
b	0	1	1	1
1	0	a	b	1

Figure 2: $(L, \to)$ demonstrating independence of modus ponens in Fact 2.1.

Proposition 3 *For any bounded lattice L and binary operation $\to$ on L, the following are equivalent:*

1. *$\to$ is a* Heyting implication, *i.e., for all $a, b, c \in L$,*

$$a \wedge b \leq c \text{ iff } a \leq b \to c;$$

2. *$\to$ is a preconditional satisfying modus ponens and weak monotonicity;*

3. *$\to$ satisfies axioms 3 and 4 of preconditionals, modus ponens, and weak monotonicity.*

Proof. The implication from 1 to 2 is straightforward and the implication from 2 to 3 is immediate.

From 3 to 1, supposing $a \leq b \to c$, we have $a \wedge b \leq (b \to c) \wedge b \leq c$ by modus ponens. Conversely, supposing $a \wedge b \leq c$, we have

$$\begin{aligned} a &\leq b \to a && \text{by weak monotonicity} \\ &\leq b \to (a \wedge b) && \text{by axiom 3 of preconditionals} \\ &\leq b \to (a \wedge b \wedge c) && \text{by our assumption that } a \wedge b \leq c \\ &\leq b \to c && \text{by axiom 4 of preconditionals.} \end{aligned}$$

$\square$

Fact 3 *Axioms 3 and 4 of preconditionals, modus ponens, and weak monotonicity are mutually independent.*

Proof. For axiom 3, we can again use the two-element lattice on $\{0, 1\}$ with $\to$ defined by $a \to b = b$, as in the proof of Fact 1.

For axiom 4, consider the three-element lattice on $\{0, 1/2, 1\}$ with $\to$ defined as follows: $x \to y = 1$ if $x = y$ and otherwise $x \to y = y$. Then $0 \to 0 = 1 \not\leq 1/2 = 0 \to 1/2$, so axiom 4 does not hold. However, one can check that axiom 3, modus ponens, and weak monotonicity hold.

For modus ponens, consider again the two-element lattice on $\{0,1\}$ with $\to$ defined by $a \to b = 1$. Since $1 \wedge (1 \to 0) = 1 \wedge 1 = 1 \not\leq 0$, modus ponens does not hold, but the other axioms clearly do.

Finally, for weak monotonicity, consider again the two-element lattice on $\{0,1\}$ with $\to$ defined by $a \to b = 0$. Since $1 \not\leq 0 = 1 \to 1$, weak monotonicity does not hold, but the other axioms clearly do. □

A natural weakening of modus ponens to consider is that the derived $\neg$ operation is a *semicomplementation*: $a \wedge (a \to 0) = 0$. Let us say that a *proto-Heyting implication* is a preconditional satisfying weak monotonicity and semicomplementation. The implication used in the proof of Fact 2.1 is a proto-Heyting implication that is not a Heyting implication.

2.4 Sasaki hook

For our second example, an *ortholattice* is a bounded lattice L equipped with a unary operation $\neg$, called an *orthocomplementation*, satisfying

- antitonicity: $a \leq b$ implies $\neg b \leq \neg a$,
- semicomplementation: $a \wedge \neg a = 0$, and
- involution: $\neg\neg a = a$.

From these properties, one can derive excluded middle $(a \vee \neg a = 1)$[2] and De Morgan's laws $(\neg(a \vee b) = \neg a \wedge \neg b$ and $\neg(a \wedge b) = \neg a \vee \neg b)$.

In an ortholattice, the *Sasaki hook* is the binary operation defined by

$$a \overset{s}{\to} b := \neg a \vee (a \wedge b) = \neg(a \wedge \neg(a \wedge b)).$$

The following is immediate from Proposition 2.

Corollary 1 *In any ortholattice, the Sasaki hook is a preconditional.*

Next we add axioms on a preconditional $\to$ to characterize the Sasaki hook. First, note that one half of the equation $a \to b = \neg a \vee (a \wedge b)$, where $\neg$ is now defined from $\to$, already follows from the preconditional axioms.

Lemma 1 *For any preconditional $\to$ on a bounded lattice, we have*

$$(a \to 0) \vee (a \wedge b) \leq a \to b.$$

---

[2]Since $a \leq a \vee \neg a$ and $\neg a \leq a \vee \neg a$, we have $\neg(a \vee \neg a) \leq \neg a \wedge \neg\neg a = 0$, so $\neg 0 \leq \neg\neg(a \vee \neg a) = a \vee \neg a$. Finally, $1 \leq \neg\neg 1$ and $\neg 1 = 1 \wedge \neg 1 = 0$, so $1 \leq \neg 0$, which with the previous sentence implies $1 \leq a \vee \neg a$.

Preconditionals

Proof. We have $a \to 0 \leq a \to b$ by axiom 4 of preconditionals and $a \wedge b \leq a \to b$ by axiom 2 of preconditionals, so $(a \to 0) \vee (a \wedge b) \leq a \to b$. □

To prove the reverse inequality, we assume that the negation defined by $\to$ is an involutive semicomplementation.

Proposition 4 *For any bounded lattice L and binary operation $\to$ on L, the following are equivalent:*

1. *$\to$ is a preconditional with $a \wedge (a \to 0) = 0$ and $(a \to 0) \to 0 = a$;*

2. *L equipped with $\neg$ defined by $\neg a := a \to 0$ is an ortholattice and $\to$ coincides with the Sasaki hook: $a \to b = \neg a \vee (a \wedge b)$.*

Proof. From 2 to 1, that $\to$ is a preconditional follows from Corollary 1. That $a \wedge (a \to 0) = 0$ and $(a \to 0) \to 0 = a$ follows from the definition of $\neg$ and the assumption that $\neg$ is an orthocomplementation.

From 1 to 2, first we show that $\neg$ is an orthocomplementation. Both $a \wedge \neg a = 0$ and $\neg \neg a = a$ follow from our assumptions on $\to$ and the definition of $\neg$. That $a \leq b$ implies $\neg b \leq \neg a$ is given by Proposition 1.1.

Finally, we show $a \to b = \neg a \vee (a \wedge b)$. The right-to-left inclusion is given by Lemma 1, so it only remains to show $a \to b \leq \neg a \vee (a \wedge b)$:

$$x \wedge \neg y \leq \neg y$$
$\Rightarrow \quad \neg \neg y \leq \neg(x \wedge \neg y) \quad$ by antitonicity for $\neg$
$\Rightarrow \quad y \leq \neg(x \wedge \neg y) \quad$ by involution for $\neg$
$\Rightarrow \quad x \to y \leq x \to \neg(x \wedge \neg y) \quad$ by axiom 4 of preconditionals
$\Rightarrow \quad x \to y \leq x \to ((x \wedge \neg y) \to 0) \quad$ by definition of $\neg$
$\Rightarrow \quad x \to y \leq (x \wedge \neg y) \to 0 \quad$ by axiom 5 of preconditionals
$\Rightarrow \quad x \to y \leq \neg(x \wedge \neg y) \quad$ by definition of $\neg$
$\Rightarrow \quad x \to y \leq \neg x \vee y \quad$ by De Morgan's law and involution for $\neg$
$\Rightarrow \quad a \to (a \wedge b) \leq \neg a \vee (a \wedge b) \quad$ substituting a for x, $a \wedge b$ for y
$\Rightarrow \quad a \to b \leq \neg a \vee (a \wedge b) \quad$ by axiom 3 of preconditionals. □

An ortholattice is *orthomodular* if $a \leq b$ implies $b = a \vee (\neg a \wedge b)$. In fact, as observed by Mittelstaedt (1972), orthomodularity is equivalent to the Sasaki hook satisfying modus ponens.

Lemma 2 (Mittelstaedt) *An ortholattice L is orthomodular if and only if $a \wedge (\neg a \vee (a \wedge b)) \leq b$ for all $a, b \in L$.*

Combining Lemma 2 with Proposition 4 yields the following.

Proposition 5 *For any bounded lattice L and binary operation $\to$ on L, the following are equivalent:*

1. *$\to$ is a preconditional satisfying modus ponens and $(a \to 0) \to 0 = a$;*

2. *L equipped with $\neg$ defined by $\neg a := a \to 0$ is an orthomodular lattice and $\to$ coincides with the Sasaki hook: $a \to b = \neg a \vee (a \wedge b)$.*

Figure 3 summarizes the relations between the classes of preconditionals covered so far (OL and OML stand for ortho- and orthomodular lattices). We also add the classical material implication of Boolean algebras, which is equivalent to Heyting implication with involution of $\neg$ and to Sasaki hook in orthomodular lattices with weak monotonicity (by Proposition 3, Lemma 2).

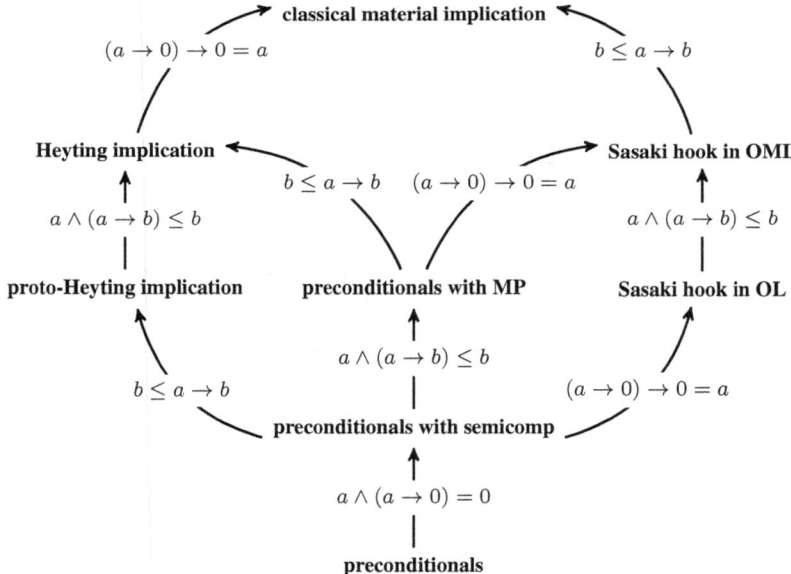

Figure 3: Classes of preconditionals.

2.5 Lewis-Stalnaker-style conditionals

The third example of lattices with preconditionals that we will consider are Boolean algebras with Lewis-Stalnaker-style conditionals (Stalnaker 1968, Lewis 1973) satisfying the axiom of *flattening* (Mandelkern 2024, §§ 6.4.1-6.4.2, citing Cian Dorr, p.c.):

$$a \to ((a \wedge b) \to c) = (a \wedge b) \to c.$$

Axiom 5 of preconditionals is simply the left-to-right inclusion.

Lewis-Stalnaker-style (set-)selection function semantics in effect treats a conditional $a \to b$ as the result of applying an a-indexed normal modal operator $\Box_a$ to b. That is, there is a binary relation R_a between worlds, and $a \to b$ is true at w iff all R_a-accessible worlds from w make b true. Further constraints are imposed so that the relations R_a can be derived from well-founded preorderings of the set of worlds: $wR_a v$ iff v is one of the closest a-worlds to w according to the well-founded preorder $\leqslant_w$ associated with w. But for our purposes here, the key aspects of the Lewis-Stalnaker (set-)selection function semantics are captured by the following definition.

Definition 2 *A* selection frame *is a pair* $(W, \{R_A\}_{A \subseteq W})$ *where W is a nonempty set and each R_A is a binary relation on W satisfying the following for all $w, v \in W$ and $A \subseteq W$:*

1. success: *if $wR_A v$, then $v \in A$;*

2. centering: *if $w \in A$, then $wR_A v$ iff $v = w$.*

Such a frame is functional *if it satisfies the following:*

3. *if $wR_A v$ and $wR_A u$, then $v = u$.*

Such a frame is strongly dense *if it satisfies the following:*

4. *if $wR_{A \cap B} v$, then $\exists u$: $wR_A u$ and $uR_{A \cap B} v$.*

Strong density says that instead of conditioning on a stronger proposition, one can first condition on a weaker one, then condition on the stronger one, and end up in the same state as one would reach by conditioning on the stronger proposition straightaway. Though not all Lewis-Stalnaker-style frames that satisfy success, centering, and functionality are strongly dense, the following are, as observed by Boylan and Mandelkern (2022).

Example 1 Given a well-ordered set $(W, \leqslant)$, for any $w \in W$ and $A \subseteq W$, let wR_Av iff v is the first world in A according to $\leqslant$ such that $w \leqslant v$. Then $(W, \{R_A\}_{A \subseteq W})$ is a strongly dense, functional selection frame.

Proposition 6 *For any strongly dense selection frame* $(W, \{R_A\}_{A \subseteq W})$, *the operation* $\to_R$ *defined by*

$$A \to_R B := \Box_A B = \{w \in W \mid \text{for all } v \in W, wR_Av \Rightarrow v \in B\}$$

is a preconditional on the Boolean algebra $\wp(W)$.

Proof. We must check the following for $\to = \to_R$:

1. $W \to A \subseteq A$; 2. $A \cap B \subseteq A \to B$; 3. $A \to B \subseteq A \to (A \cap B)$;
4. $A \to (B \cap C) \subseteq A \to B$; 5. $A \to ((A \cap B) \to C) \subseteq (A \cap B) \to C$.

Condition 1 follows from centering (in particular, the right-to-left direction of the biconditional in centering), as does condition 2 (but now the left-to-right direction of the biconditional in centering); condition 3 follows from success; and condition 4 is immediate from the definition of $\to$. For condition 5, suppose $w \in A \to ((A \cap B) \to C)$ and $wR_{A \cap B}v$. Then by strong density, there is a u such that wR_Au and $uR_{A \cap B}v$. Since $w \in A \to ((A \cap B) \to C)$ and wR_Au, we have $u \in (A \cap B) \to C$, which with $uR_{A \cap B}v$ yields $v \in C$. This shows that $w \in (A \cap B) \to C$. □

The key principles validated by selection frames beyond the axioms of preconditionals are modus ponens,

- identity: $a \to a = 1$, and
- normality: $(a \to b) \wedge (a \to c) \leq a \to (b \wedge c)$.

Functional frames also validate

- negation import: $\neg(a \to b) \leq a \to \neg b$.

Proposition 7 *Let B be a finite Boolean algebra equipped with a preconditional $\to$ satisfying modus ponens, identity, normality, and negation import. Let W be the set of atoms of B and $\widehat{(\cdot)}$ the isomorphism from B to $\wp(W)$. For $a \in B$, define*

$$wR_{\widehat{a}}v \text{ iff for all } b \in B, w \leq a \to b \text{ implies } v \leq b.$$

Then $(W, \{R_A\}_{A \subseteq W})$ is a strongly dense, functional selection frame, and $(B, \to)$ is isomorphic to $(\wp(W), \to_R)$.

Preconditionals

Proof. First, we check the following:
1. success: if $wR_{\widehat{a}}v$, then $v \leq a$;
2. centering: if $w \leq a$, then $wR_{\widehat{a}}v$ iff $v = w$;
3. functionality: if $wR_{\widehat{a}}v$ and $wR_{\widehat{a}}u$, then $v = u$;
4. strong density: if $wR_{\widehat{a \wedge b}}v$, then $\exists u\colon wR_{\widehat{a}}u$ and $uR_{\widehat{a \wedge b}}v$.

For success, given $w \leq a \to a$ from identity, $wR_{\widehat{a}}v$ implies $v \leq a$.

For centering, assume $w \leq a$. Modus ponens for $\to$ yields $wR_{\widehat{a}}w$. Then the rest of centering follows given functionality, which we prove next.

For functionality, if $wR_{\widehat{a}}v$, then we claim $w \leq a \to v$. For if $w \not\leq a \to v$, then since w is an atom, we have $w \leq \neg(a \to v)$ and hence $w \leq a \to \neg v$ by negation import, contradicting $wR_{\widehat{a}}v$. Then since $w \leq a \to v$, if $wR_{\widehat{a}}u$, then $u \leq v$, which implies $u = v$ given that v is an atom.

For strong density, assume $wR_{\widehat{a \wedge b}}v$. Let

$$x = \bigwedge\{y \in B \mid w \leq a \to y\}.$$

We claim that $x \neq 0$. Otherwise there are $y_1, \ldots, y_n$ such that

$$w \leq (a \to y_1) \wedge \cdots \wedge (a \to y_n) \text{ and } y_1 \wedge \cdots \wedge y_n = 0.$$

But then by normality and axioms 4-5 of preconditionals,

$$w \leq a \to (y_1 \wedge \cdots \wedge y_n) = a \to 0 \leq (a \wedge b) \to 0,$$

contradicting $wR_{\widehat{a \wedge b}}v$. Hence $x \neq 0$, so there is an atom $u \leq x$, and by construction of x, $wR_{\widehat{a}}u$. To show $uR_{\widehat{a \wedge b}}v$, suppose $u \leq (a \wedge b) \to c$. Then $w \leq a \to ((a \wedge b) \to c)$, for otherwise $w \leq a \to \neg((a \wedge b) \to c)$ using negation import, in which case $u \leq \neg((a \wedge b) \to c)$ by construction of x, which contradicts $u \leq (a \wedge b) \to c$ given modus ponens. Then since $w \leq a \to ((a \wedge b) \to c)$, we have $w \leq (a \wedge b) \to c$ by axiom 5 of preconditionals, which with $wR_{\widehat{a \wedge b}}v$ implies $v \leq c$. Thus, $uR_{\widehat{a \wedge b}}v$.

Finally, the proof that $\widehat{a \to b} = \widehat{a} \to_R \widehat{b}$ is just like the usual proof for a normal modal box. □

This kind of result can be generalized beyond finite algebras (e.g., to complete and atomic algebras, assuming $\bigwedge\{a \to b_i \mid i \in I\} \leq a \to \bigwedge\{b_i \mid i \in I\}$) and beyond Boolean algebras, but we will not do so here.

Finally, let us return to the Heyting and Sasaki examples of Sections 2.3 and 2.4, respectively, with identity, normality, and negation import in mind.

Wesley H. Holliday

Proposition 8

1. *Heyting implications satisfy identity, normality, and negation import.*

2. *Proto-Heyting implications satisfy identity and negation import, but not necessarily normality.*[3]

3. *In orthomodular lattices, Sasaki hook satisfies normality, but not necessarily negation import.*

4. *In ortholattices, Sasaki hook satisfies identity but not necessarily normality.*[4]

Proof. Part 1 is standard. For part 2 and identity, by weak monotonicity and axiom 3 of preconditionals, $1 \leq a \to 1 \leq a \to (a \wedge 1) = a \to a$. For negation import, by weak monotonicity, $b \leq a \to b$, so $\neg(a \to b) \leq \neg b$ by Lemma 1.1, and $\neg b \leq a \to \neg b$ by weak monotonicity again, so indeed $\neg(a \to b) \leq a \to \neg b$. For a proto-Heyting implication that does not satisfy normality, consider the lattice on the left of Figure 4, and define $\to$ such that for any elements x and y: $1 \to x = x$; $x \to 0 = 0$ if $x \neq 0$; $0 \to 0 = 1$; $x \to y = 1$ if $x \neq 1$, $y \neq 0$, and $(x, y) \neq (a, d)$; and $a \to d = d$. Then $(a \to b) \wedge (a \to c) = 1 \wedge 1 = 1$ but $a \to (b \wedge c) = a \to d = d$, so normality does not hold, but one can check that the other axioms hold.

For part 3, in an orthomodular lattice, $\neg x \leq y$ implies $y \leq \neg x \vee (x \wedge y)$. Then since $\neg a \leq (a \to b) \wedge (a \to c)$, we have

$$\begin{aligned}(a \to b) \wedge (a \to c) &\leq \neg a \vee (a \wedge (a \to b) \wedge (a \to c)) \\ &\leq \neg a \vee (a \wedge b \wedge c) \quad \text{by Lemma 2} \\ &= a \to (b \wedge c) \quad \text{by definition of Sasaki hook.}\end{aligned}$$

To see that negation import is not necessarily satisfied, consider the modular lattice M4 with elements $\{0, a, b, c, d, 1\}$ such that a, b, c, d are incomparable in the lattice order, turned into an ortholattice with $\neg a = b$ and $\neg c = d$. Then for the Sasaki hook we have $\neg(a \to c) = \neg(\neg a \vee (a \wedge c)) = \neg(\neg a \vee 0) = \neg\neg a = a$, whereas $a \to \neg c = \neg a \vee (a \wedge \neg c) = \neg a \vee (a \wedge d) = \neg a \vee 0 = \neg a$.

For part 4, identity for Sasaki hook is just excluded middle. For a failure of normality, consider the ortholattice in Figure 4. Then for the Sasaki hook,

---

[3]Moreover, there are normal proto-Heyting implications that are not Heyting implications, such as the implication used in the proof of Fact 2.1.

[4]Moreover, there are non-orthomodular lattices in which the Sasaki hook satisfies normality, such as the lattice O6 (the "benzene ring").

Preconditionals

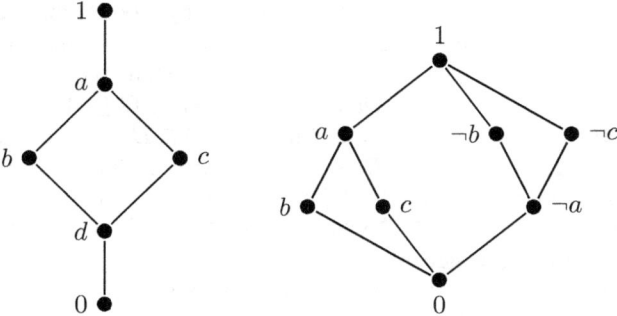

Figure 4: Left: lattice for the proof of Proposition 8.2. Right: an ortholattice in which the Sasaki hook violates normality for Proposition 8.4.

$(a \to b) \wedge (a \to c) = (\neg a \vee (a \wedge b)) \wedge (\neg a \vee (a \wedge c)) = (\neg a \vee b) \wedge (\neg a \vee c) = 1 \wedge 1 = 1$, whereas $a \to (b \wedge c) = \neg a \vee (a \wedge b \wedge c) = \neg a \vee 0 = \neg a$. □

Though the conditionals in Proposition 8 are normal if they satisfy modus ponens, this is not the case for preconditionals in general.[5] An instructive example comes from the following probabilistic interpretation. Given $W = \{0, \dots, 10\}$, define for each $w \in W$ a measure $\mu_w \colon \wp(W) \to [0,1]$ by $\mu_w(\{w\}) = .9$, $\mu_w(\{v\}) = .01$ for $v \neq w$, and $\mu_w(A) = \sum_{v \in A} \mu_w(\{v\})$ for non-singleton $A \subseteq W$. Then for $A, B \subseteq W$ with $A \neq \varnothing$, let

$$A \to B = \{w \in W : \mu_w(B \mid A) \geq .9\},$$

where as usual $\mu_w(B \mid A) = \mu_w(A \cap B)/\mu_w(A)$, and $\varnothing \to B = W$.

Proposition 9 *The operation $\to$ just defined is a preconditional on $\wp(W)$ satisfying modus ponens but not normality.*

Proof. First observe that $A \cap (A \to B) \subseteq B$, because if $w \in A$, then since $\mu_w(\{w\}) = .9$, we can have $\mu_w(B \mid A) \geq .9$ only if $w \in B$. This also shows that axiom 1 of preconditionals holds. For axiom 2, if $w \in A \cap B$, then again since $\mu_w(\{w\}) = .9$, we have $\mu_w(B \mid A) \geq .9$, so $w \in A \to B$. Axioms 3 and 4 also clearly hold. For axiom 5, suppose $w \in A \to ((A \cap B) \to C)$,

---

[5] Note that preconditionals satisfying normality but not modus ponens can be obtained from the examples in Footnotes 3 and 4.

so $\mu_w((A \cap B) \to C \mid A) \geq .9$. If $w \in A$, then by modus ponens, we have $w \in (A \cap B) \to C$, as desired. So suppose $w \not\in A$. Further suppose for contradiction that $w \not\in (A \cap B) \to C$, i.e., $\mu_w(A \cap B \cap C)/\mu_w(A \cap B) < .9$, so $|A \cap B \cap C| < |A \cap B|$. We claim that $(A \cap B) \to C \subseteq A \cap B$. Consider any $x \in W \setminus (A \cap B)$ with $x \neq w$. Since $x \not\in A \cap B$, $w \not\in A \cap B$, $x \neq w$, and $|W| = 11$, we have $|A \cap B| \leq 9$, which with $|A \cap B \cap C| < |A \cap B|$ implies $\mu_x(C \cap A \cap B)/\mu_x(A \cap B) \leq .08/.09 < .9$, so $x \not\in (A \cap B) \to C$. Thus, $(A \cap B) \to C \subseteq A \cap B$, which implies $(A \cap B) \to C \subseteq A \cap B \cap C$ by modus ponens, which with $\mu_w((A \cap B) \to C \mid A) \geq .9$ implies $\mu_w(A \cap B \cap C \mid A) \geq .9$, which in turn implies $\mu_w(C \mid A \cap B) \geq .9$ and hence $w \in (A \cap B) \to C$.

Finally, for the failure of normality, $0 \in \{1, \ldots, 10\} \to \{1, \ldots, 9\}$ and $0 \in \{1, \ldots, 10\} \to \{2, \ldots, 10\}$, but $0 \not\in \{1, \ldots, 10\} \to \{2, \ldots, 9\}$. □

3 Relational representation of lattices with preconditionals

Having hopefully shown the interest of the class of preconditionals from an axiomatic perspective, let us return to the semantic origin of preconditionals. Given a set X, binary relation $\lhd$ on X (let $\rhd := \lhd^{-1}$), and $A, B \subseteq X$, define[6]

$$A \to_\lhd B = \{x \in X \mid \forall y \lhd x \, (y \in A \Rightarrow \exists z \rhd y : z \in A \cap B)\}, \quad (1)$$

As shown in Holliday 2023, the operation $c_\lhd$ defined by $c_\lhd(A) = X \to_\lhd A$ is a closure operator, so its fixpoints ordered by inclusion form a complete lattice $\mathfrak{L}(X, \lhd)$ with meet as intersection and join as $\bigvee_\lhd \{A_i \mid i \in I\} = c_\lhd(\bigcup\{A_i \mid i \in I\})$. If $A, B \in \mathfrak{L}(X, \lhd)$, then $A \to_\lhd B \in \mathfrak{L}(X, \lhd)$,[7] so we may regard $\to_\lhd$ as an operation on $\mathfrak{L}(X, \lhd)$. The following is easy to check.

Fact 4 *The operation $\to_\lhd$ on $\mathfrak{L}(X, \lhd)$ is a preconditional.*

Example 2 In the relational frame shown at the top of Figure 5, the arrow from y to x indicates $x \lhd y$, etc. Reflexive loops are assumed but not shown. Transitive arrows are *not* assumed. Note that $y \in G$ but $y \not\in B \to_\lhd G$, contra weak monotonicity. Also note that $y \in (P \to_\lhd 0) \to_\lhd 0 = 0 \to_\lhd 0 = 1$, but $y \not\in P$, contra involution. So $\to_\lhd$ is neither Heyting nor Sasaki.

---

[6]In Holliday 2023, we denoted the operation defined in (1) by '$\twoheadrightarrow_\lhd$' in order to distinguish it from a different operation denoted by '$\to_\lhd$'. Since we do not need that distinction here, we will use the cleaner '$\to_\lhd$' for the operation defined in (1).

[7]Obviously $A \to_\lhd B \subseteq X \to_\lhd (A \to_\lhd B)$. For the reverse, suppose $x \not\in A \to_\lhd B$. Hence there is some $y \lhd x$ such that $y \in A$ and for all $z \rhd y$, we have $z \not\in A \cap B$. It follows

Preconditionals

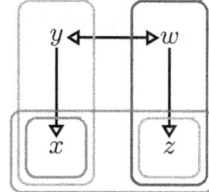

$R = \{x\}$
$O = \{z\}$
$G = \{x, y\}$
$B = \{w, z\}$
$P = \{x, z\}$

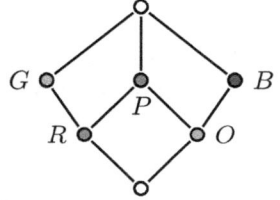

$\to_\triangleleft$	0	R	O	G	P	B	1
0	1	1	1	1	1	1	1
R	B	1	B	1	1	B	1
O	G	G	1	G	1	1	1
G	O	P	O	1	P	O	1
P	0	G	B	G	1	B	1
B	R	R	P	R	P	1	1
1	0	R	O	G	P	B	1

Figure 5: A relational frame (top) giving rise to a lattice with preconditional (bottom).

In fact, every complete lattice with a preconditional can be represented by such a frame $(X, \triangleleft)$. More generally, we have the following.

Theorem 1 (Holliday 2023, Theorem 6.3) *Let L be a bounded lattice and $\to$ a preconditional on L. Then where*

$$P = \{(x, x \to y) \mid x, y \in L\} \text{ and } (a, b) \triangleleft (c, d) \text{ if } c \not\leq b,$$

there is a complete embedding of $(L, \to)$ into $(\mathfrak{L}(P, \triangleleft), \to_\triangleleft)$, which is an isomorphism if L is complete.

Let us take this a step further with a topological representation. Important precedents for the non-conditional aspects of this representation can be found in Urquhart 1978, Allwein and Hartonas 1993, Ploščica 1995, and Craig, Haviar, and Priestley 2013. Given a bounded lattice L and a preconditional $\to$, define $\mathsf{Fl}(L, \to) = (X, \triangleleft)$ as follows: X is the set of all pairs (F, I) such that F is a filter in L, I is an ideal in L, and for all $a, b \in L$:

$$\text{if } a \in F \text{ and } a \wedge b \in I, \text{ then } a \to b \in I.$$

---

that there is no $w \triangleright y$ with $w \in A \to_\triangleleft B$. Thus, if $x \notin A \to_\triangleleft B$, then there is a $y \triangleleft x$ such that for all $w \triangleright y$, $w \notin A \to_\triangleleft B$, which shows that $x \notin X \to_\triangleleft (A \to_\triangleleft B)$.

Call such an (F, I) a *consonant* filter-ideal pair. We define $(F, I) \triangleleft (F', I')$ if $I \cap F' = \varnothing$. Finally, given $a \in L$, let $\widehat{a} = \{(F, I) \in X \mid a \in F\}$, and let $\mathsf{S}(L, \to)$ be $\mathsf{Fl}(L, \to)$ endowed with the topology generated by $\{\widehat{a} \mid a \in L\}$ (cf. Bezhanishvili and Holliday 2020).

Theorem 2 *For any bounded lattice L and preconditional $\to$ on L, the map $a \mapsto \widehat{a}$ is*

1. *an embedding of $(L, \to)$ into $(\mathfrak{L}(\mathsf{Fl}(L, \to)), \to_\triangleleft)$ and*

2. *an isomorphism from L to the subalgebra of $(\mathfrak{L}(\mathsf{Fl}(L, \to)), \to_\triangleleft)$ consisting of elements of $\mathfrak{L}(\mathsf{Fl}(L, \to))$ that are compact open in the space $\mathsf{S}(L, \to)$.*

Proof. For $a \in L$, let $\uparrow a$ (resp. $\downarrow a$) be the principal filter (resp. ideal) generated by a. First we claim that for any $a, b \in L$, $(\uparrow a, \downarrow a \to b) \in X$. For suppose $c \in \uparrow a$ and $c \wedge d \in \downarrow a \to b$, so $a \le c$ and $c \wedge d \le a \to b$. Then

$$\begin{aligned}
c \to d &\le c \to (c \wedge d) \text{ by axiom 3 of preconditionals}\\
&\le c \to (a \to b) \text{ by axiom 4, since } c \wedge d \le a \to b\\
&= c \to ((a \wedge c) \to b) \text{ since } a \le c\\
&\le (a \wedge c) \to b \text{ by axiom 5}\\
&= a \to b \text{ since } a \le c,
\end{aligned}$$

so $c \to d \in \downarrow a \to b$. Since $b = 1 \to b$ by axioms 1 and 2 of preconditionals, it follows that $(\uparrow 1, \downarrow b) = (\uparrow 1, \downarrow 1 \to b) \in X$ as well.

Using the above facts, the proof that $\widehat{a}$ belongs to $\mathfrak{L}(\mathsf{Fl}(L, \to))$ and that $a \mapsto \widehat{a}$ is injective and preserves $\wedge$ and $\vee$ is the same as in the proof of Theorem 4.30.1 in Holliday 2023. Also note that $\widehat{1} = X$ and $\widehat{0} = c_\triangleleft(\varnothing)$.

Next we show that $\widehat{a \to b} = \widehat{a} \to_\triangleleft \widehat{b}$. First suppose $(F, I) \in \widehat{a \to b}$, $(F', I') \triangleleft (F, I)$, and $(F', I') \in \widehat{a}$, so $a \in F'$. Since $(F, I) \in \widehat{a \to b}$, we have $a \to b \in F$, which with $(F', I') \triangleleft (F, I)$ implies $a \to b \notin I'$, which with $a \in F'$ and the definition of X implies $a \wedge b \notin I'$. Now let $F'' = \uparrow a \wedge b$ and $I'' = \downarrow (a \wedge b) \to 0$. Then $(F'', I'') \in X$, $(F', I') \triangleleft (F'', I'')$, and $(F'', I'') \in \widehat{a \wedge b}$. Thus, $(F, I) \in \widehat{a} \to_\triangleleft \widehat{b}$. Conversely, if $(F, I) \notin \widehat{a \to b}$, so $a \to b \notin F$, then setting $(F', I') = (\uparrow a, \downarrow a \to b)$, we have $(F', I') \in X$ and $(F', I') \triangleleft (F, I)$. Now consider any (F'', I'') such that $(F', I') \triangleleft (F'', I'')$, so $a \to b \notin F''$. Then by axiom 2 of preconditionals, $a \wedge b \notin F''$, so $(F'', I'') \notin \widehat{a \wedge b}$. Thus, $(F, I) \notin \widehat{a} \to_\triangleleft \widehat{b}$.

The proof of part 2 is the same as the proof of Theorem 4.30.2 in Holliday 2023. □

Finally, we can characterize the spaces equipped with a relation $\triangleleft$ that are isomorphic to $\mathsf{S}(L, \rightarrow)$ for some $(L, \rightarrow)$. Let X be a topological space and $\triangleleft$ a binary relation on X. Let $\mathsf{COFix}(X, \triangleleft)$ be the set of all compact open sets of X that are also fixpoints of $c_\triangleleft$. If $\mathsf{COFix}(X, \triangleleft)$ is a lattice with meet as $\cap$ and join as $\vee_\triangleleft$, F is a filter in this lattice, and I is an ideal, then we can speak of (F, I) being a consonant filter-ideal pair as defined above Theorem 2, using $\rightarrow_\triangleleft$ in the definition. Given $x \in X$, let

$$\mathsf{F}(x) = \{U \in \mathsf{COFix}(X, \triangleleft) \mid x \in U\}$$
$$\mathsf{I}(x) = \{U \in \mathsf{COFix}(X, \triangleleft) \mid \forall y \triangleright x \ y \notin U\}$$

and note that $(\mathsf{F}(x), \mathsf{I}(x))$ is a consonant filter-ideal pair from $\mathsf{COFix}(X, \triangleleft)$. For if $U \in \mathsf{F}(x)$ and $U \cap V \in \mathsf{I}(x)$, then $x \in U$ but for all $y \triangleright x$, $y \notin U \cap V$, which implies that for all $y \triangleright x$, $y \notin U \rightarrow_\triangleleft V$, so $U \rightarrow_\triangleleft V \in \mathsf{I}(x)$.

Proposition 10 *For any space X and binary relation $\triangleleft$ on X, there is a bounded lattice L with preconditional $\rightarrow$ such that $(X, \triangleleft)$ and $\mathsf{S}(L, \rightarrow)$ are homeomorphic as spaces and isomorphic as relational frames iff the following conditions hold for all $x, y \in X$:*

1. $x = y$ *iff* $(\mathsf{F}(x), \mathsf{I}(x)) = (\mathsf{F}(y), \mathsf{I}(y))$;

2. $\mathsf{COFix}(X, \triangleleft)$ *contains X and $c_\triangleleft(\varnothing)$, is closed under $\cap$, $\vee_\triangleleft$, and $\rightarrow_\triangleleft$, and forms a basis for X;*

3. *each consonant filter-ideal pair from $\mathsf{COFix}(X, \triangleleft)$ is $(\mathsf{F}(x), \mathsf{I}(x))$ for some $x \in X$;*

4. $x \triangleleft y$ *iff* $\mathsf{I}(x) \cap \mathsf{F}(y) = \varnothing$.

The proof is very similar to that of Proposition 3.21 in Holliday 2022, but we provide the adapted proof here for convenience.

Proof. Suppose there is such an L. It suffices to show $\mathsf{S}(L, \rightarrow)$ satisfies conditions 1–4 in place of $(X, \triangleleft)$. That condition 2 holds for $\mathsf{COFix}(\mathsf{S}(L, \rightarrow))$ and $\mathsf{S}(L, \rightarrow)$ follows from the proof of Theorem 2. Let φ be the isomorphism $a \mapsto \widehat{a}$ from L to $\mathsf{COFix}(\mathsf{S}(L, \rightarrow))$ in Theorem 2, which induces a bijection $(F, I) \mapsto (\varphi[F], \varphi[I])$ between consonant filter-ideal pairs of L and of $\mathsf{COFix}(\mathsf{S}(L, \to))$. Conditions 1, 3, and 4 follow from the fact that

$$\text{for any } x = (F, I) \in \mathsf{S}(L, \to), (\varphi[F], \varphi[I]) = (\mathsf{F}(x), \mathsf{I}(x)). \tag{2}$$

First, $\widehat{a} \in \varphi[F]$ iff $a \in F$ iff $x \in \widehat{a}$ iff $\widehat{a} \in \mathsf{F}(x)$. Second, $\widehat{a} \in \varphi[I]$ iff $a \in I$, and we claim that $a \in I$ iff $\widehat{a} \in \mathsf{I}(x)$, i.e., for all $(F', I') \rhd (F, I)$, $(F', I') \not\ni \widehat{a}$, i.e., $a \not\in F'$. If $a \in I$ and $(F, I) \lhd (F', I')$, then $a \not\in F'$ by definition of $\lhd$. Conversely, if $a \not\in I$, let $F' = {\uparrow}a$ and $I' = {\downarrow}a \to 0$, so (F', I') is consonant by the proof of Theorem 2, $(F, I) \lhd (F', I')$ since $a \not\in I$, and $a \in F'$. Thus, $\widehat{a} \not\in \mathsf{I}(x)$. This completes the proof of (2).

Now for condition 1, given $x, y \in \mathsf{S}(L, \to)$ with $x = (F, I)$ and $y = (F', I')$, we have $(F, I) = (F', I')$ iff $(\varphi[F], \varphi[I]) = (\varphi[F'], \varphi[I'])$ iff $(\mathsf{F}(x), \mathsf{I}(x)) = (\mathsf{F}(y), \mathsf{I}(y))$; similarly, for condition 4, $(F, I) \lhd (F', I')$ iff $I \cap F' = \varnothing$ iff $\varphi[I] \cap \varphi[F'] = \varnothing$ iff $\mathsf{I}(x) \cap \mathsf{F}(y) = \varnothing$. Finally, for condition 3, if $(\mathcal{F}, \mathcal{I})$ is a consonant filter-ideal pair from $\mathsf{COFix}(\mathsf{S}(L, \to))$, then setting $x = (\varphi^{-1}[\mathcal{F}], \varphi^{-1}[\mathcal{I}])$, we have $x \in \mathsf{S}(L, \to)$ and $(\mathcal{F}, \mathcal{I}) = (\mathsf{F}(x), \mathsf{I}(x))$.

Assuming X satisfies the conditions, $\mathsf{COFix}(X, \lhd)$ is a bounded lattice with a preconditional by Fact 4, and we define a map ϵ from $(X, \lhd)$ to $\mathsf{S}(\mathsf{COFix}(X, \lhd), \to_\lhd)$ by $\epsilon(x) = (\mathsf{F}(x), \mathsf{I}(x))$. The proof that ϵ is a homeomorphism using conditions 1–3 is analogous to the proof of Theorem 5.4(2) in Bezhanishvili and Holliday 2020. That ϵ preserves and reflects $\lhd$ follows from condition 4. □

4 Conclusion

We have seen that preconditionals encompass several familiar classes of conditionals, including Heyting implication, the Sasaki hook, and Lewis-Stalnaker-style conditionals satisfying flattening. Lattices with these implications are therefore covered by the general representation in Theorem 2. A natural next step is to try to obtain nice characterizations of the relational-topological duals of these kinds of algebras, as well as of more novel kinds—such as lattices with normal preconditionals—not to mention going beyond representation to categorical duality. We hope that the delineation of preconditionals and their relational-topological representation may help to provide a unified view of a vast landscape of conditionals arising in logic.

References

Allwein, G., & Hartonas, C. (1993). *Duality for bounded lattices.* (Indiana University Logic Group, Preprint Series, IULG-93-25 (1993))

Bezhanishvili, N., & Holliday, W. H. (2020). Choice-free Stone duality. *The Journal of Symbolic Logic, 85*(1), 109-148.

Boylan, D., & Mandelkern, M. (2022). *Logic and information sensitivity.* (ESSLLI 2022 lecture notes, https://mandelkern.hosting.nyu.edu/ESSLLI2022Day3.pdf)

Craig, A. P. K., Haviar, M., & Priestley, H. A. (2013). A fresh perspective on canonical extensions for bounded lattices. *Applied Categorical Structures, 21,* 725-749.

Egré, P., & Rott, H. (2021). *The Logic of Conditionals.* https://plato.stanford.edu/archives/win2021/entries/logic-conditionals/. (In *The Stanford Encyclopedia of Philosophy* (Winter 2021 Edition))

Holliday, W. H. (2022). Compatibility and accessibility: lattice representations for semantics of non-classical and modal logics. In D. F. Duque & A. Palmigiano & S. Pinchinat (Eds.), *Advances in Modal Logic, Vol. 14* (pp. 507-529). London: College Publications.

Holliday, W. H. (2023). A fundamental non-classical logic. *Logics, 1,* 36-79.

Lewis, D. (1973). *Counterfactuals.* Oxford: Basil Blackwell.

Mandelkern, M. (2024). *Bounded Meaning: The Dynamics of Interpretation.* Oxford: Oxford University Press.

Mittelstaedt, P. (1972). On the interpretation of the lattice of subspaces of Hilbert space as a propositional calculus. *Zeitschrift für Naturforschung, 27a,* 1358-1362.

Ploščica, M. (1995). A natural representation of bounded lattices. *Tatra Mountains Mathematical Publication, 5,* 75-88.

Stalnaker, R. C. (1968). A theory of conditionals. In N. Rescher (Ed.), *Studies in Logical Theory* (pp. 98–112). Oxford: Blackwell.

Urquhart, A. (1978). A topological representation theory for lattices. *Algebra Universalis, 8,* 45-58.

Wesley H. Holliday
University of California, Berkeley
USA
E-mail: wesholliday@berkeley.edu

Aristotle's Syllogistic Logic as a Theory of Arithmetical Kind

LADISLAV KVASZ[1]

Abstract: There is a paradox associated with Aristotle's logic. On the one hand, the theory of syllogisms is generally considered to be the first system of formal logic in history, but on the other hand, this logic is not used by ancient scholars such as Euclid, Archimedes or Ptolemy, nor is it used by Aristotle himself in his writings on natural sciences. The aim of this paper is to try to explain this paradox through an analysis of the epistemological structure of the language in which Aristotle's logic is formulated. In the first two sections, I introduce the notions of relational, compositional and deductive synthesis and of phenomenal, ontological and causal reduction. On the basis of these notions, I will then distinguish three kinds of theories – theories of the physical kind, theories of the mathematical kind, and theories of the arithmetical kind. I will try to show that syllogistic logic is a theory of the arithmetical kind and therefore cannot be used in the analysis of the logical structure of mathematical and physical theories. If this interpretation is correct, it shows why the creators of modern science like Galileo or Descartes first had to reject the framework of Aristotelian logic and the methodology based on it.

Keywords: syllogistic logic, compositional synthesis, idealization, elementary arithmetic, theories of the arithmetical kind

1 Introduction

In the history of logic, we encounter a remarkable paradox. On the one hand, Aristotle, thanks to his syllogistics, is considered the founder of formal logic. Jan Łukasiewicz writes: "the introduction of variables into logic is one of Aristotle's greatest inventions. ... Through the use of variables,

---

[1] I thank Vladimír Balek, Matyáš Havrda, Palo Labuda, Róbert Maco and Peter Volek for stimulating discussions on Aristotle, his science and method. This paper was written as part of the research activities of the Centre for Science, Technology and Society Studies of the Institute of Philosophy of the Czech Academy of Sciences in Prague.

Aristotle became the creator of formal logic" (Łukasiewicz, 1957, pp. 7–8). John Corcoran sums up the importance of syllogistics by saying: "In presenting axiomatic science, it has been customary to leave the underlying logic implicit. Neither in Euclid's geometry nor in Hilbert's do we find a codification of the logical rules used in deriving theorems from axioms. ... We argue that in those chapters of the *First Analytics*, Aristotle developed a logical theory that included a theory of deduction for deriving categorical conclusions from categorical premises. We further argue that Aristotle regarded the logic thus developed as the underlying logic of the axiomatic disciplines discussed in the first chapter of the *Second Analytics*." (Corcoran, 1974, pp. 89–90).

On the other hand, neither Aristotle in his writings on natural science nor scholars such as Euclid, Archimedes, or Ptolemy used syllogistic logic. Ian Mueller comes to the following conclusion: "In Euclid we find no awareness of syllogistics or even of the basic idea of logic that the validity of an argument depends on its form. Aristotle's references to mathematics are either to general points about deductive reasoning or, insofar as they refer specifically to syllogistics, are erroneous because they are based only on syllogistics and not on an independent analysis of mathematical proofs" (Mueller, 1974, p. 37). He concludes his article by saying, "Aristotle's formulation of the syllogistic is essentially independent of Greek mathematics. There is no evidence that he or his peripatetic followers studied mathematical proofs in detail." (Mueller, 1974, p. 66).

Jonathan Barnes puts this paradox as follows: "The method that Aristotle employs in his scientific and philosophical writings and the method that he prescribes for scientific and philosophical activity in the *Second Analytics* are clearly not the same. ... If the Organon were lost, we would have no reason to suppose that Aristotle discovered the syllogisms and was immensely proud of it. This, then, is the Problem: on the one hand, a highly formalized theory of scientific methodology; on the other, a practice that is not at all affected by formalization and itself exhibits rich and varied methodological aspirations: how can the two be reconciled?" (Barnes, 1969, pp. 123–125). Euclid does not use syllogistic logic. And not because mathematicians are biased against the use of it. Even if they wanted to, syllogistic logic cannot be used in most mathematical proofs.

Łukasiewicz claims that Aristotle does not admit singular terms in syllogisms (Łukasiewicz, 1957, p. 1). Mueller remarks on this as follows: "If Łukasiewicz is right, then none of Euclid's arguments is an Aristotelian

Aristotle's Syllogistic Logic as a Theory of Arithmetical Kind

syllogism" (Mueller, 1974, p. 67).[2] The reason is simple. A Euclidean proof has six parts: 1. *protasis* (stating the proposition in general terms); 2. *ecthesis* (relating the proposition to a specific situation represented by a figure); 3. *diorismos* (formulating the problem in the context of that situation); 4. *kataskeyé* (constructing additional elements of the figure); 5. *apodeixis* (a proof in which both the elements adduced in the ekthesis and the elements constructed in the kataskeyé are used); and 6. *symperasma* (reformulating the result obtained in a specific situation into a general form). These six parts are illustrated by Mueller in the proof of the first proposition of the first book of the *Elements*. For my purposes, I will restrict myself to the first three. *Protasis*: "Construct an equilateral triangle over a given straight line." *Ecthesis*: "Let AB be a given straight line." *Diorismos*: "Over the line AB, then, construct an equilateral triangle" (quoted from Šír, 2011, p. 119).

The aim of this paper is to explain the difference between the *general character* of argumentation in Aristotelian logic and the *particular character* of argumentation in Euclidean mathematics. I will try to clarify where this difference comes from and what its significance is. Answering these questions will help explain why Aristotle's syllogistic cannot be used in mathematics and why Aristotle could not rely on the practice of mathematical proof in constructing his syllogistic. I will try to show that Aristotle's syllogistic logic is *a theory of the arithmetical kind*,[3] which makes it fundamentally different from the theories contained in Euclid's *Elements*, which are *theories of the mathematical kind*. That Aristotle's logic is a theory of the arithmetical kind is not to be understood in the sense that this logic constitutes an implicit (or underlying) logic of elementary arithmetic, since already elementary arithmetical relations such as $3 + 2 = 5$ have no natural transcription into the subject–predicate form. Nor do I understand the proofs in Aristotle's

---

[2]This statement refers to Aristotle's version of syllogistic logic. The development of logic in the Middle Ages seems to have overcome this limitation and made it possible to introduce singular terms into syllogisms (see Novák & Dvořák, 2007, pp. 115–118). A modern interpretation of syllogistics can be found in (Kolman & Punčochář, 2015), and a philosophical clarification of its assumptions is given by Jaroslav Peregrin and Marta Vlasáková in their book *Philosophy of Logic* (Peregrin & Vlasáková, 2017).

[3]The word *arithmetic* and its derivatives will be understood in a narrow sense as referring to practical counting, which we know from ancient Egypt, Babylon and other ancient civilizations. I will distinguish it from the number theory we know from ancient Greece. As Gisela Striker writes: "The word *syllogismos* as a technical term was not an invention of Aristotle. It is derived from the verb *syllogizestai*, which in common Greek means to compose or count" (Striker, 2009, p. 79). This is only a clue, not an argument, but it indicates the direction I want to take in my argument.

syllogistic to be modeled on arithmetical proofs, as they are later in Frege.[4] Rather, I mean that the way the language of Aristotle's logic constructs representations of complex notions and situations (by adding attributes to the relevant definitions) is analogous to the way the language of arithmetic constructs representations of large sets (namely, by adding units). Thus, I do not claim that Aristotle investigated the logic of arithmetic, nor that he imitated it, but I claim only that Aristotle's logic and elementary arithmetic have an analogous linguistic structure. My thesis, then, is not that Aristotle derived his logic from arithmetic (as Frege later derived his logic from the analysis of mathematical proofs in theoretical arithmetic). Aristotle probably developed the syllogistics from an analysis of dialogical argumentation. However, the way he *formalized* this dialectical argumentation used not a mathematical structure but the structure of the language of elementary arithmetic. He may have adopted the use of schematic letters from the mathematics of his time, but what he used these letters for has nothing to do with mathematics. He used them to express contents that are of the arithmetical kind.

Even a cursory glance reveals that the relations captured by syllogistics are relations between classes of objects (as in counting) and not between individual objects, which appear in Euclid's proofs. When I claim that seven (ants) is more than five (elephants), the relation is between the number of elements of two classes, and not between the individual objects that belong to those classes (ants and elephants). In answering the question, "which is more, seven ants or five elephants?", I have to ignore the nature of the elements (i.e., the fact that one elephant is greater than a million ants), stay at the level of the classes (concepts), and answer the question based on the relationship of the number of elements in each class. In mathematics, on the other hand, although the goal is to prove a general proposition, the proposition is often so subtle that it cannot be proved at the general level of concepts (or classes) and one must go to the level of individual objects and rely on their structure for the proof. Aristotle's syllogistics, therefore, by remaining at the general level of concepts, is closer to calculus than to mathematics.

---

[4]The relation of Frege's logic to Aristotle's is laid out in Vojtěch Kolman's book *Filosofie čísla*, where he writes: "If we understand logic as an enterprise concerned with questions of justification of the validity of judgments, ... Aristotle's syllogistics represents an important precedent, which Frege 'only' modified and developed in a certain way. The fact that in practice we are unable to derive this extension of it from traditional logic, then, is simply due to the fact that Frege, unlike Aristotle, was oriented towards a different discourse of judgment and therefore had to start *ab ovo*" (Kolman, 2008, p.163). The aim of my essay is to describe the difference between these discourses more precisely.

Aristotle's Syllogistic Logic as a Theory of Arithmetical Kind

Mueller points out an interesting fact: "In the systematic presentation of categorical syllogisms in the first twenty-two chapters of the *First Analytics*, Aristotle nowhere refers to mathematics. His examples are always of the type 'white', 'man', 'beast', and suggest a close relationship between Aristotle's logic and the somewhat puzzling dialectical activities associated with Plato's Academy. The difficulty of reconciling mathematical arguments with syllogistic form may explain the absence of mathematical references in these chapters" (Mueller, 1974, p. 48). In what follows, I will try to explain where these difficulties in reconciling mathematical arguments and syllogistics stem from and what exactly they consist of.

2 The concept of relational, compositional and deductive synthesis of language

Language contributes to the formulation of scientific theories by, among other things, enabling the connection of what would otherwise remain unconnected. In this context, I will speak of the synthetic role of language and propose to distinguish three kinds of synthesis. The first kind of synthesis I will call ***relational synthesis***, and I understand it as the ability of language to relate different aspects of phenomena to each other. For example, Kepler's third law gives into a direct proportionality relation the third power of the length of the main semi-axis of the planet's orbit a^3 and the square of the planet's orbital period around the Sun T^2. This relationship is inaccessible to senses because we cannot perceive the square of time. However, the language of algebra allows us to construct a quantity T^2 from the observed quantity T and relate it to a^3. Thus, the language of algebra allows us to relate two aspects of the planet's motion (a, T) that, although we can observe, we are unable to relate at the empirical level. Therefore, I propose to call this ability of the language relational synthesis.

In addition to relational synthesis, I propose to introduce ***compositional synthesis***, by which I mean the ability of language to produce representations of complex systems. For example, Newtonian mechanics can represent the simultaneous motion of several bodies, whereas Aristotelian physics could only describe the motion of a single body. Aristotelian physics could not combine the description of motion of individual bodies into a unified dynamical system. Newtonian physics is able to do this because it assigns a state description (i.e., a linguistic representation) to each body, and from the state descriptions of individual bodies it creates a state description of the entire system. The above difference in the description of motion shows that the language of Newtonian physics has (unlike the language of Aristotelian physics) a compositional synthesis; it allows one to compose motions.

However, we need to be more precise when introducing the concept of compositional synthesis. Compositional synthesis of the language of Newtonian physics involves, in addition to the description of the state of each body by its position and momentum, the description of the action between bodies by means of forces and the description of the change of state induced by this action. The language of Aristotelian physics cannot describe the state of moving bodies, the action of forces, or the change of state induced by that action. We express this fact by saying that the language of Aristotle's physics does not have a compositional synthesis of the physical type, it does not have the means to combine the motions of bodies into a common dynamic. Aristotle's physics has a compositional synthesis of the mathematical type. Aristotle (like Ptolemy, Copernicus, and Galileo) composes the motions of individual bodies in the same way that Euclidean geometry, it combines lines and circles into a common geometric configuration: it assigns positions to individual elements relative to other elements. In contrast, Newton composes the motions of individual bodies by making them into a common dynamic system in which the individual bodies interact. When I say of Aristotle's physics that it is a theory without a compositional synthesis, I mean without a compositional synthesis of the physical kind that would make it possible to combine the motions of bodies into a unified dynamical system. Aristotle's physics unifies the elements of the universe into a common geometric structure, not into a unified physical dynamic.

The third kind of synthesis I propose to call ***deductive synthesis***, and by it, I mean the ability of language to allow us to draw conclusions analytically from representations of reality. The language of Newtonian mechanics allows us to infer the future dynamical state of a system from the knowledge of its present state and the of forces acting on it, i.e., it allows us to 'compute' the motion, so to speak.

The introduction of relational, compositional and deductive synthesis allows the thesis that Aristotle's logic is a theory of the arithmetical kind to be divided into two parts. The first part of the thesis is that *arithmetic and mathematics have a fundamentally different relational, compositional, and deductive synthesis*. It can be said that the relational synthesis of the language of arithmetic makes it possible to put numbers obtained by counting into the relation of being less than, equal to, or greater than. However, two numbers

Aristotle's Syllogistic Logic as a Theory of Arithmetical Kind

cannot be similar to each other, they cannot intersect, nor can one number be perpendicular to the other. Thus, the language of arithmetic has a more limited relational synthesis than the language of geometry, in which there is a relation of similarity between two figures (a constant ratio of the lengths of their corresponding parts) and two lines can be in different positions relative to each other. The compositional synthesis of the language of arithmetic is trivial – a number is a collection of units, so *a unit enters a number only by its presence*. In contrast, the compositional synthesis of Euclid's *Elements* is given by postulates. Straight lines and circles enter the construction of a composite geometric figure not only by their presence, but in many different ways. The deductive synthesis of arithmetic is given by the rules of addition, multiplication, subtraction and division, while the deductive synthesis of Euclid's *Elements* is given by the axioms.[5] Axioms, such as "What is equal to the same is equal to each other" (Šír, 2011, p. 117), legitimize the steps of mathematical proofs. The second part of the thesis is that *the relational, compositional and deductive synthesis of Aristotle's logic is identical to the relational, compositional and deductive synthesis of elementary arithmetic.*[6]

These parts require detailed justification. However, it is already clear at first sight that the subject-predicate structure of the propositions of Aristotle's logic is related to the compositional synthesis of the arithmetical kind. Subjects of a certain species enter the higher genus in Aristotelian predication in the same way as units enter the number in arithmetic: by their presence. If this interpretation is correct, it also answers the question why Aristotle did not use syllogistics in his *Physics*, *Metaphysics*, or *Natural History*, and why Euclid did not use it in the *Elements*. The reason is that the relations Aristotle and Euclid described in these works are more complex than a language with such a simple relational, compositional, and deductive synthesis as the language of arithmetic can express.

Aristotle often uses the example of a theorem that Euclid formulates in the words: "In every triangle, the three angles inside the triangle are equal to two right angles" (Šír, 2011, p. 125), while Aristotle puts it in the form "Every triangle has two right angles". At first glance, this may seem to be a

---

[5] I understand axioms in the strict sense that distinguishes between axioms and postulates.

[6] It is important to emphasize that this thesis does not state the aims of Aristotle's project in the *Second Analytics*, but speaks exclusively of the linguistic framework of syllogistic logic. As Barnes has noted, "to a surprising degree the *Second Analytics* is independent of the theory of syllogisms developed in the *First Analytics*: whatever the explanation for this fact, it has the happy consequence that Aristotle's theory of axiomatization does not inherit all the shortcomings of his theory of inference" (Barnes, 1982, p. 53).

reformulation of the original mathematical theorem into the form of a general categorical judgment required by syllogistic. The subject is a triangle and the predicate are two right angles. However, as Mueller notes, "in the proof, the terms triangle and two right angles cannot act as categorical terms because the proof consists in dividing the triangle and the two right angles into parts, the spatial relations of these parts being decisive" (Mueller, 1974, p. 52). This illustrates the different synthesis of the language of Aristotle's syllogistics, which uses the indivisible terms "triangle" and "two right angles," and the language of Euclid's *Elements*, which moves from general terms to specific objects in order to divide them into elements (points, segments, and arcs of circles), put them into relations with each other, add more elements, and then use these relations to prove the original theorem.

3 The concept of phenomenal, ontological and causal reduction

The relational, compositional, and deductive synthesis we have just described, did not arise on their own. They arose in the wake of three kinds of reduction, which I propose to call phenomenal, ontological, and causal reduction. Thus, in addition to relational, compositional, and deductive synthesis, language also contributes to the formulation of scientific theories by allowing us to *reduce* the richness of phenomena, the complexity of structures, and the intricacies of influences in ways that allow us to better understand reality.

Consider the relational synthesis, which I illustrated using Kepler's third law. In order to relate the square of the orbital period T^2 and the third power of the length of the major semi-axis of the orbit of the planet a^3, the orbital period and the major semi-axis first had to be quantified, i.e., *reduced to a number*. Physics has similarly reduced many other phenomena, from velocity and acceleration through temperature and atmospheric pressure to color. That this is indeed a reduction can be illustrated by temperature, which are given at the phenomenal level as experiences of coldness, pleasant warmth, or painful heat. Physics reduces these phenomena to numbers, to 41°C. Since this is a reduction of a phenomenon to its linguistic representation (which in the case of physics is a quantity), I propose to call the transition from phenomenal content to linguistic representation a *phenomenal reduction*.

Compositional synthesis, which I have illustrated with the example of the ability of language of physics to combine the motions of bodies into a common dynamical system, can be understood in a similar way. Again, for

Aristotle's Syllogistic Logic as a Theory of Arithmetical Kind

the language of physics to be able to combine the motions of bodies into a common dynamical system, it was first necessary to reduce the description of individual bodies to their state. In the case of Newtonian mechanics, the state is given by position and velocity (or more precisely, vector of position and vector of momentum). Since only objects having a certain ontological status can have a state, I propose to call this kind of reduction an ***ontological reduction***.[7]

Finally, let us consider the deductive synthesis, which I have illustrated by the ability to infer the future state of a dynamical system from knowledge of its state and the forces acting on it. Again, in order for the language to relate the current state of a physical system to its future state (by means of a differential equation), all interactions must be reduced to the action of forces, that can then be inserted into the equation of motion. I propose to call this kind of reduction a ***causal reduction***.[8]

The notions of phenomenal, ontological and causal reduction allow us to understand why the relational, compositional and deductive synthesis of the language of Aristotle's logic is arithmetical. Consider the propositions:

(1) Bodies attract (each other). (2) Birds fly.

In both cases, a non-essential attribute is ascribed to the subject, and therefore, from the point of view of Aristotle's logic, these propositions are analogous, so that they could function as premises of syllogistic inference. However, the similarity between propositions (1) and (2) is illusory. From the point of view of Newtonian physics, there is a fundamental difference between these propositions.

The first proposition is a popular formulation of the law of gravity. The law of gravity is an idealized law according to which two bodies of masses m_1 and m_2 located at distance r attract each other by a force F whose magnitude

---

[7] I have illustrated the ontological reduction at the language of physics. Nevertheless, ontological reduction is also used in mathematics. In mathematics, compositional synthesis does not consist in unifying moving bodies into a dynamical system, but in uniting elements in the process of construction of a complex structure. Therefore, the ontological reduction of the mathematical kind has the form of the reduction of various shapes to elementary figures (combinations of points, lines and circles in geometry) which may have different relative positions (analogy of different states of the individual bodies in mechanics). It is the specification of relative positions that allows the elementary figures of a construction to be combined to form a complex geometric representation.

[8] Unlike phenomenal and ontological reduction, only the language of physics has causal reduction. Mathematics and arithmetic do not use action in the description of phenomena, and therefore do not have causal reduction.

is equal to $F = \kappa \cdot (m_1 \cdot m_2)/r^2$. In the law of gravity, the quantities m_1 and m_2 are the result of a phenomenal reduction that replaces every aspect of reality with a physical quantity. The distance r of the bodies is the length of the difference of their position vectors, so it is part of the description of the state of the bodies, i.e. it is the result of the ontological reduction that reduces the description of bodies to their position and momentum. Finally, the force F on the left-hand side of the law of gravity is the result of a causal reduction that reduces all interaction to the action of forces. The reduction of phenomena to magnitudes; of motions to states; and of interactions to forces embeds the law of gravity into a framework in which we can explain phenomena, such as tides, the motion of comets, or the irregularities in the motion of the moon, from the fact that "bodies attract".

On the contrary, there is no idealized law behind the proposition "Birds fly". Therefore, no non-trivial implications can be drawn from it. We must have known that geese fly when we classified them as birds. From the fact that birds fly, nothing comparable to the tides can be inferred, because the statement lacks a linguistic framework similar to that provided to the law of gravity by Newtonian mechanics, a framework with a non-trivial relational, compositional and deductive synthesis of the physical kind.

4 Idealization as a change of the relational, compositional and deductive synthesis

I introduced six concepts: relational, compositional and deductive synthesis, and phenomenal, ontological and causal reduction. Let us now turn to a description of the linguistic framework of which these concepts are part. This is the *framework of idealization*. By *idealization I mean a linguistic reduction of a particular fragment of reality*. We use ordinary language (among other things) to describe reality. That is, we take reality as it appears to us in ordinary experience, and use language to describe its various aspects (for example, we use numbers to describe the quantitative aspect of reality). Sometimes we succeed in constructing such a *linguistic representation of the aspect of reality* under investigation that it allows us to find out practically everything relevant about that aspect. And not only that. Sometimes it is possible to build into the representation rules for manipulating linguistic expressions, so that by means of these manipulations (in the case of numbers, we call these rules counting) we can predict the outcomes of real events and answer various questions. The linguistic representation thus begins to function as a kind of parallel reality alongside the factual reality. However, reality and its representation can come into conflict. An example of such a conflict was incommensurability, discovered in ancient Greece. *Idealization* is a response to such a conflict and it consists in making the rules for manipulating linguistic representations self-sustained and using linguistic representations as if they were independent of the relevant aspect of reality (thus, in response to the discovery of incommensurability, postulates and axioms were explicitly formulated).

There have been three idealizations so far. The first was the *idealization of number* that took place more than 4000 years ago in the ancient civilizations of Egypt, Mesopotamia, India, China, and others. It has resulted in numbers whose ideality lies in the sharpness, precision and unambiguity of the rules for manipulating them, and also in the fact that the number series transcends all the limits of empirical experience. The linguistic practice associated with numbers we call counting, and the doctrine that systematizes the rules of this linguistic practice we call elementary arithmetic.

The second idealization was the *idealization of shape*, which took place in ancient Greece in the 4th century BC (more than 1500 years after the first idealization). It resulted in idealized geometric figures, and the linguistic practice associated with it consists in constructing geometric figures and proving propositions about them. In Euclid, this practice absorbed arithmetic, and turned it into number theory, since operations with numbers can be represented by manipulations of line segments.

The third idealization is the *idealization of motion*, and the linguistic practice that produced it is the practice of measurement and experimentation and the description of underlying processes using differential equations as we know them from mechanics, electrodynamics, or quantum mechanics. Before describing idealization of number, which was crucial to the emergence of Aristotle's logic, let us first describe the idealization of motion and the structure of the language of physics based on it.

5   The structure of the languages of theories of the physical kind

The idealization of motion can be divided into six steps: three linguistic reductions and three linguistic syntheses. I have already described the individual syntheses and reductions; now I will explain how they are related. The first step towards idealization of motion was the *reduction of phenom-*

ena of ordinary experience to mathematical quantities, such as for example the reduction of the sensation of heat to temperature or the reduction of the perception of color to wavelength. I will call the replacement of a phenomenon by its linguistic representation a ***phenomenal reduction***. As a rule, this is done by means of instruments such as a thermometer, barometer, spectroscope, voltmeter, etc.

The reduction of phenomena to quantities allows the emergence of ***relational synthesis***, which represents the second step of the idealization process. For example, when the passage of time is reduced to the quantity t, the language of mathematics offers t^2, t^3, t^4, ... , $\sqrt{t}$, $\sqrt[3]{t}$, $\sqrt[4]{t}$, ..., $\sin(t)$, $\log(t)$, e^t, ... Thus, the temporal aspect of a phenomenon, because of its reduction to a mathematical quantity, can enter into relationships with other aspects of that phenomenon in many different ways that are inaccessible at the phenomenal level. One manifestation of the relational synthesis of the language of mechanics is Galileo's law of free fall, according to which the path s traveled by a falling body is proportional to the square of time: $s = \frac{1}{2}gt^2$. We must realize that we have gone far beyond the limits of ordinary experience. In ordinary experience we understand what time is, we know how long lasts a second (it is approximately the time between two heartbeats), an hour, a month or a year. But what is a square hour? In the case of distance, we have perceptual access to square meters because space appears to us in three independent dimensions. But in the case of time? Can anyone imagine a square second? And how many square seconds does a square minute have? How do we know it has passed? And precisely such square seconds are proportional to distance, according to Galileo's law.

Galilean physics was based on phenomenal reduction and relational synthesis. Although Galileo achieved important results in describing free fall, projectile motion, and motion on an inclined plane, it is not hard to see that these examples hide a fundamental problem. They are examples in which only one body moves. This shows that *the language of Galilean physics lacks compositional synthesis* – the ability to unite motions of bodies into a single dynamical system. The unification of the motion of bodies into a common dynamic is preceded by the introduction of a state, that I propose to call ontological reduction. Thus, the third step of idealization of motion is the ***ontological reduction of the system of bodies to a representation of their state***. Just as phenomenal reduction replaces phenomena (e.g., the sensation of heat) by a linguistic representation using a mathematical quantity, ontological reduction replaces the being of a physical system by its linguistic representation, which is called the state of the physical system. From the

Aristotle's Syllogistic Logic as a Theory of Arithmetical Kind

description of the states of individual bodies, the state of the system as a whole can be constructed, and thus the united motion of several bodies can be described. The language of physics thus again transcends the description of phenomena. A state is an ontological category, a state can have only something that has the ontological status of being.

Just as phenomenal reduction opened up the possibility of introduction of relational synthesis, the notion of state opens up the possibility of introduction of *compositional synthesis*. In physics, compositional synthesis originally meant the ability of the language to combine the motions of bodies into a unified dynamical system. The first attempt to build compositional synthesis into the language of physics were Descartes' collision rules. The description of a collision of two bodies is a description of a composite system, because in the collision the bodies exchange part of their momentum, thus change their state of motion. Descartes thus introduced a mathematical description of interaction by contact and created the first physical language with compositional synthesis.

The third type of reduction, which together with the previous two forms the framework for the mathematical description of motion, is *causal reduction*. In consisted in Newton's replacement of the description of interaction by contact by a mathematical representation of forces acting at a distance. This reduction consists, just like the previous two cases, in replacing the verbal description of an aspect of reality by its mathematical representation. Unlike the phenomenal reduction, however, the forces acting at a contact, that Newton replaced with their mathematical representation, are not perceptually accessible.

As in the previous two cases, also causal reduction by replacing an aspect of reality by its mathematical representation opened up the possibility to introduce into language a new kind of synthesis – the *deductive synthesis*. In the case of the language of physics the deductive synthesis is based on a differential equation – the equation of motion. That equation describes the changes of the state of the system caused by action of forces over an infinitesimal time interval, and thus makes it possible to derive from the present state of the system its state at a later moment of time.

The language of Newtonian physics, by means of (1) *phenomenal reduction*, replaces phenomena by numbers. Phenomenal reduction allows the introduction of (2) *relational synthesis*, through which the various quantities are related. Certain combinations of these quantities allow (3) *ontological reduction* of motion to a description of changes of state. The introduction of a state allows the introduction of (4) *compositional synthesis*, which from the description of the state of the individual bodies constructs the state of the dynamical system that contains them. The notion of a dynamical system requires a description of the interactions between bodies, leading to (5) a *causal reduction* of all kinds of interaction to forces acting at a distance. The introduction of forces allows the temporal evolution of the state of the system under the action of forces to be calculated by means of (6) *deductive synthesis* using the equation of motion (Newton's second law).

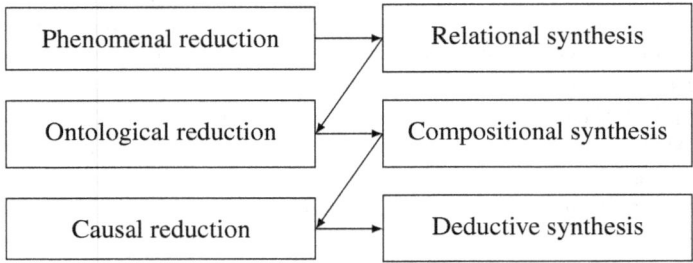

I will not analyze further the idealization of motion. I have introduced it only to clarify how the three reductions and their corresponding three syntheses are related.

6 The structure of the languages of theories of the mathematical kind

Let's move on to the idealization of shape. ***Phenomenal reduction*** consists in replacing real shapes by ideal shapes. Phenomenal reduction is followed by ***relational synthesis***. The fundamental geometric relation is the relation of similarity, which allows geometric figures to be embedded in a network of relations whose richness and subtlety exceeds those accessible to ordinary perception. An example illustrating the relational synthesis of the language of geometry is the Pythagorean theorem, according to which the squares above the hypotenuse of a right triangle have the same total area as the square above the hypotenuse. This relation is inaccessible to perception, but the language of geometry can express it.

Geometric figures enter into a series of relations that are so varied that compositional synthesis cannot be governed solely by how the figures appear, i.e. it cannot be linked directly to relational synthesis. ***Ontological reduction*** thus leads us from how a figure appears to what it actually is. Geometric shapes, understood as ideal phenomena, are replaced by geometric figures, i.e. ideal objects. The intersection of two lines, which appears as a small quadrilateral, becomes in the ontological reduction a dimensionless point.

The *compositional synthesis* of the language of geometry is based on a geometric construction whose steps are unambiguously determined by postulates. However, for such a construction to work, an ontological reduction had to take place first. Only objects can touch, intersect and enter into different relations. After the ontological reduction, the geometrician must answer the question of what is the intersection of two lines by saying that it is a point without dimensions and parts (as required by the definition), even though he is looking at two lines that intersect in a small area of a certain size and shape. Only points without parts and dimension can enter the steps of compositional synthesis: only such points can be uniquely connected by a straight line, or a circle can be circumscribed around them. Ontological reduction reduces geometric shapes so that the subsequent compositional synthesis can be unambiguous. If a point had parts, we wouldn't know which one to stick a compass into or draw a straight line through. Euclid's postulates are the principles of compositional synthesis and guarantee the unambiguity of the steps of geometric construction.

Compositional synthesis in physics was followed by causal reduction. However, the language of geometry is characterized by being a-temporal (geometric relations are timeless), and therefore we cannot speak of causal relations among geometric objects. We consider the fact that the language of mathematics does not contain causal reduction as binding for the interpretation of idealization in mathematics. The *deductive synthesis* of the language of geometry, which takes the form of discursive argumentation, is associated with the compositional synthesis without an intermediate step of causal reduction, as it is in the case the language of physics. A proof is a sequence of steps of deductive reasoning that proceed from axioms. The steps of the proof are based on the steps of construction in which new elements were gradually added to the constructed figure. They are added in specific relations to the already constructed elements, and these relations must be considered in the proof. *The language of mathematics has no causal reduction*, and so the deductive synthesis is directly connected to the compositional synthesis. This is consistent with the practice in Euclid, where the proof (*apodeixis*, using the deductive synthesis) follows the steps of construction (*kataskeyé*, i.e. compositional synthesis).

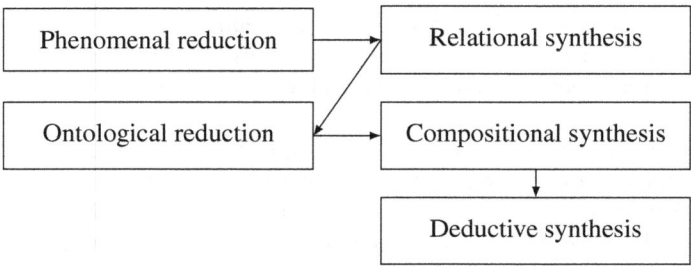

7 The structure of the languages of theories of the arithmetical kind

In the idealization of numbers, the phenomenon involved in the *phenomenal reduction* is number. Phenomenal reduction consists in reducing the objects we count to units, that is, in considering them as the same despite their apparent differences. No matter how large the objects we count, the number three is the same, whether it is three ants or three elephants. When we count shoes, no matter their size, color, or appearance, we reduce them to units. When we count, we go through the set and recite the number series. The number we end up with is the number of items we counted. Phenomenal reduction allows us to introduce *relational synthesis* into the language. The fundamental relation that enters the universe of counted things is the equality of number and the derived relation "more than".

Just as we noted in describing the idealization of shape that in geometry there is no causal reduction and the deductive synthesis is directly connected to compositional synthesis, it seems that elementary arithmetic has no ontological reduction either. That is, in arithmetic we do not refer to numbers as entities with independent existence, but leave them at the level of phenomena. The *compositional synthesis*[9] of the language of arithmetic is thus connected directly to its relational synthesis and consists in addition, subtraction, multiplication and division. The multiplication of two numbers (i.e. the act of compositional synthesis) can always be reduced to repeated additions of the unit. By adding a unit, we are able to generate all numbers regardless of any constraints, thus exceeding all quantitative limits. The structure of

---

[9]Compositional synthesis is the way a language creates complex representations (large numbers, composite figures, or systems of bodies) from elementary elements (*units* in arithmetic, lines and circles in geometry, bodies and forces in mechanics). Thus, in different languages, compositional synthesis differs as well.

the language of arithmetic is simpler than that of mathematics because *the language of arithmetic has no ontological reduction.*

The language of elementary arithmetic contains no causal reduction either. In the exposition of the language of elementary arithmetic, we can thus move directly to deductive synthesis. The result of virtually any problem concerning quantity can be solved by manipulating number symbols according to explicit rules (i.e. by means of the linguistic representation of the problem).

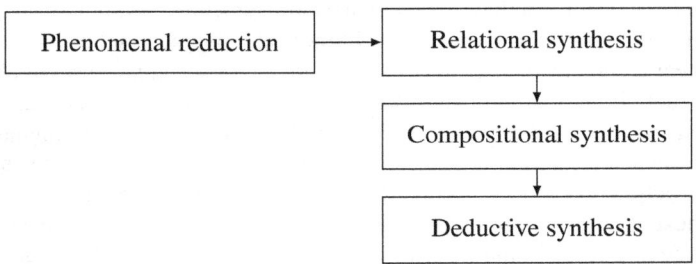

The language of elementary arithmetic is thus connected to reality only by phenomenal reduction. My thesis is that *the language of Aristotle's syllogistics is a language of the same kind.*

8 Syllogistic logic as a theory of the arithmetical kind

Thus far I introduced relational, compositional, and deductive synthesis and phenomenal, ontological and causal reduction. Now I want to show that the language of Aristotelian logic has arithmetical relational, compositional and deductive synthesis. That is, in syllogistic logic, compositional synthesis follows directly the relational synthesis (without mediating ontological reduction) and deductive synthesis follows directly the compositional synthesis (without mediating causal reduction). The only contact of the language of Aristotle's logic with reality is phenomenal reduction (i.e. predication). In counting, we view objects as bearing certain features. Depending on the particular features we choose in determining the unit (we can, for example, count blue-eyed children), we count an object according to whether or not it has the chosen features (i.e., whether it is blue-eyed and a child). That is, the phenomenal reduction consists in reducing an object to a unit. In doing so, we do not assume that the unit or the objects we count have any internal structure (unless some feature refers to such a structure), or rather, we are not concerned with any structure when counting. The number is the number of units. *A unit participates in a number only by its presence.* When counting the edges of a cube, it does not matter that some are perpendicular to each other and others are parallel. All that matters for counting is that there are twelve of them.

When Aristotle chose the subject-predicate construction of the judgment as the basis of logic, he was proceeding in accordance with the character of the phenomenal reduction of the language of elementary arithmetic, i.e., as if the universe of discourse were an arithmetical universe, a set of individuals that bear certain features (and not a geometrical universe, i.e., a set of objects constructed from straight lines and circles; or a physical universe, i.e., a system of moving bodies interacting with each other). The language of Aristotle's logic assumes no mathematical compositional synthesis (i.e., construction) for the elements of its universe. Therefore, the deductive synthesis that his logic is able to express is the deductive synthesis of a language of the arithmetical kind. This is why Aristotle's logic does not allow one to analyze geometric proofs. Such proofs follow the compositional synthesis of a construction and draw consequences from its step. The objects to which Aristotle's logic refers to by means of concepts are understood as bearers of various features. Hence the synthesis of Aristotle's logic (compared to geometry) is trivial, and therefore the propositions it can prove are trivial as well.

Of course, an *entire construction*, such as the construction of an isosceles right-angled triangle, can be declared as a *single feature* and in this way the *predicate* "isosceles right-angled triangle" can be created. But by turning an entire geometric construction into a single feature, we lose access to its particular *steps*, as well as to the *elements* that were introduced in these steps, and to the *relations* in which the construction puts these elements. A mathematical proof relies on such elements and their relations. Therefore, by turning a construction into a feature, we lose access not only to the elements of compositional synthesis, but also to the steps of deductive synthesis. We are left only with the triviality that every isosceles right-angled triangle is right-angled and isosceles. Using syllogistic logic, we cannot prove that a circle inscribed in an isosceles right-angled triangle divides its sides in the ratio $(\sqrt{2} - 1) : 1$. What we need in order to prove this is not a middle term that would be the predicate or subject of the premise of the syllogism, but some auxiliary objects whose addition to the triangle would make it possible to deduce the ratio. What we are looking for is a decomposition of

Aristotle's Syllogistic Logic as a Theory of Arithmetical Kind

the *object* into steps of construction, not a decomposition of the *proposition* into premises of a syllogism.

Aristotle illustrates his method with the example of a lunar eclipse. However, being eclipsed by the Earth is not a property of the Moon, but of the configuration formed by the Moon, the Earth, and the Sun. The language used to explain the eclipse must be able to reconstruct the geometric configuration of these three bodies. It must therefore be a language with a compositional synthesis of the language of geometry and not of the language of arithmetic. The Earth, the Moon and the Sun must enter into the configuration together with their spatial relations and not merely by their presence. It is indeed possible to declare the entire structure of relations to be a feature of a single element at any step of a geometric proof. For example, the sum of the angles in a triangle can be declared to be a feature of the triangle. However, by doing this we lose the ability to prove the statement that the sum of the angles in a triangle is 180° because we lose access to the elements that appear in the proof.

8.1 The relational synthesis of the language of Aristotelian logic

The relations by which the language of elementary arithmetic enriches the relations accessible to ordinary experience consist in comparing the number of elements of larger sets, where we can no longer make do with estimation and have to start counting. These are the "less" and "more" relations. Aristotle's understanding of judgment as a relation of subject and predicate seems to be similar in character to the relations of less and more. The judgments of syllogistic logic can be translated into arithmetical relations. That is, the richness of the relations produced by the relational synthesis of Aristotelian logic is not of the mathematical kind, but rather of the arithmetical one.

That every dog is a mammal can be seen as an (approximate, distant, but still) parallel to counting dogs when counting mammals (because dogs are mammals). I do not claim that this correspondence is exact or universal, nor that it allows us to reduce Aristotle's logic to counting. I do not claim that it is an identity, but only a similarity. It seems that the degree of complexity of the arithmetical relations (which is the expression of the relational synthesis of the language of arithmetic) and the degree of complexity of the propositions of Aristotle's logic, are similar. When I count dogs, I take a feature (an aspect, a property, a predicate) and looking at a given object I examine whether it is a bearer of that feature (and then I count it) or it is not a bearer of it (and so I don't count it). When I verify the judgment that all dogs are mammals,

I search the universe, and when I come across a dog, I look to see if it is a mammal. The relational synthesis of elementary arithmetic and Aristotle's logic are similar: in both, objects enter the situation only by their presence.

The relational synthesis of the language of mathematics is different. When I consider the claim that in a right-angled triangle the square above the hypotenuse is equal to the sum of the squares above the other two sides, I don't go through the individual right-angled triangles and look to see if they have the feature "the square above the hypotenuse is equal to the sum of the squares above the other two sides". First, this property is not obvious (it cannot be grasped by means of phenomenal reduction), and second, it is not the point of the theorem. The sentence is a proposition in which the sides of the triangle, the squares above them, and indeed the whole configuration that Euclid gives in his proof, enter into relations with each other. Of course, we can try to give this statement the form required by Aristotelian logic by declaring the right-angled triangle to be the subject, and the feature "the square above the hypotenuse is equal to the sum of the squares above the two other sides" to be the predicate. But that doesn't help us. We will never construct a syllogism in which some third term is added so as to get the statement of the Pythagorean theorem as a conclusion from two valid premises. Proving in geometry doesn't work that way. First of all, we have to add the height on the hypotenuse to the figure of the triangle. This height divides the square above the hypotenuse into two rectangles, and we have to split the proof into two parts – in one we prove the equality of one of these rectangles with the square above one side, and in the other the equality of the other rectangle with the square above the other side. This means that we add auxiliary elements to the object. We decompose the property "the square above the hypotenuse is equal to the sum of the squares above the other two sides" into two properties, namely "the square above one side is equal to the rectangle with sides c and c_1" and "the square above the other side is equal to the rectangle with sides c and c_2", where c_1 and c_2 are the segments into which the height divided the hypotenuse.

8.2 The compositional synthesis of the language of Aristotelian logic

Elementary arithmetic differs from mathematics also in its compositional synthesis, i.e. the way it creates representations of complex situations. In the case of arithmetic, the basic element that emerges from phenomenal reduction is the unit, and the compositional synthesis of the language of arithmetic consists in the repeated addition of the unit. *The units participate in the number only by their presence*, the units have no relative position, no distinction is made between cases of their participation. In geometry, on the other hand, the basic elements that emerge from the ontological reduction are points, straight lines and circles. The compositional synthesis of the language of synthetic geometry is given by Euclid's postulates. These determine how complex formations emerge from elementary objects (points, straight lines and circles) in the process of construction. Elementary objects can have different relative positions (lines can be parallel or they can intersect; a circle can touch a line, intersect it, or lie outside it), and the steps of construction must consider the relative positions of the elements.

My thesis is that the compositional synthesis of the language of Aristotle's logic is akin to the compositional synthesis of the language of elementary arithmetic. When Aristotle pronounces the judgment "Every horse is brown", he reduces the horse to something akin to a unit. The horse participates in the set of brown things in the same way that a counted thing participates in a number, that is, *by its presence*. By contrast, when a mathematician defines a concept, that concept is given not as a set of features, but as a construction.

8.3 The deductive synthesis of the language of Aristotelian logic

Just as I have argued that the relational and compositional syntheses of Aristotle's logic are arithmetical in nature, so I claim that the deductive synthesis of Aristotle's logic is analogous to the deductive synthesis of elementary arithmetic. That is, an Aristotelian syllogism resembles a rule of computation. If every horse is a mammal and every mammal is warm-blooded, we can infer that every horse is warm-blooded in the same way that we infer that horses are included when counting mammals and that mammals are included when counting warm-blooded animals. Thus, the Barbara rule is arithmetic in nature.

9 Conclusion

Finally, I want to emphasize that the fact that Aristotle's logic has a relational, compositional, and deductive synthesis analogous to the syntheses of the language of elementary arithmetic is not a shortcoming, and I do not see my analysis as a criticism of Aristotelian logic. On the contrary, the fact that Frege modeled his logic after the formulaic language of arithmetic,[10]

---

[10] The subtitle of the book (Frege, 1879) is "a formula language, modeled upon that of arithmetic, for pure thought".

i.e., like Aristotle, and he only used a stronger mathematical apparatus (the notion of function) at the technical level, shows that Aristotle found a path leading to mathematical logic.

That Aristotle's logic arose in connection with elementary arithmetic should not be surprising. It may be that in order to be able to reflect on the underlying logic of a certain layer of idealization (in this case, the idealization of number), we need to get "beyond" it – we need to have at our disposal the resources of the next layer of idealization (in Aristotle's case, the idealization of shape). Frege seems to have been in a similar situation. He was able to reflect the logic of mathematical proofs by using the notion of function. The notion of function was introduced in the course of the idealization of motion. It was only the linguistic resources created in the course of idealization of motion, which proved rich enough to enable Frege to describe the logic of a mathematical proof. Aristotle's situation seems to have been analogous. The linguistic resources created in the course of idealization of shape (for example, the use of letters in the description of geometrical figures or the notion of a diagram – in Greek schema) were rich enough to allow Aristotle to create the first ever system of logic. The fact that this was a system related to the logic of elementary arithmetic may be due to the fact that it forms a much simpler, but nevertheless, a consistent, coherent and complete system of logical inference.

References

Barnes, J. (1969). Aristotle's theory of demonstration. *Phronesis*, *14*(2), 123–152.
Barnes, J. (1982). *Aristotle*. New York: Oxford University Press.
Corcoran, J. (1974). Aristotle's natural deduction system. In J. Corcoran (Ed.), *Ancient Logic and its Modern Interpretations* (pp. 85–131). Dordrecht: D. Reidel.
Frege, G. (1879). *Begriffsschrift, eine der arithmetischen nachgebildete Formelsprache des reinen Denkens*. Hildesheim: Georg Olms.
Kolman, V. (2008). *Filosofie čísla*. Prague: Filosofia.
Kolman, V., & Punčochář, V. (2015). *Formy jazyka*. Prague: Filosofia.
Łukasiewicz, J. (1957). *Aristotle's Syllogistic from the Standpoint of Modern Formal Logic*. Oxford: Oxford University Press.

Aristotle's Syllogistic Logic as a Theory of Arithmetical Kind

Mueller, I. (1974). Greek mathematics and Greek logic. In J. Corcoran (Ed.), *Ancient Logic and its Modern Interpretations* (pp. 35–70). Dordrecht: D. Reidel.
Novák, L., & Dvořák, P. (2007). *Úvod do logiky aristotelovské tradice*. České Budějovice: Teologická fakulta Jihočeské Univerzity.
Peregrin, J., & Vlasáková, M. (2017). *Filosofie logiky*. Prague: Filosofia.
Striker, G. (Ed.). (2009). *Aristotle's* Prior Analytics *Book I*. Oxford: Clarendon Press. (Translated with an introduction and commentary)
Šír, Z. (2011). *Řecké matematické texty*. Prague: Oikoymenh. (A bilingual Greek and Czech edition)

Ladislav Kvasz
Czech Academy of Sciences, Institute of Philosophy
The Czech Republic
E-mail: kvasz@flu.cas.cz

Dialectical Dispositions and Logic

LIONEL SHAPIRO[1]

Abstract: According to *dialectical disposition expressivism* about conjunction, disjunction, and negation, the function of these connectives is to convey dispositions speakers have with respect to challenging and meeting challenges to assertions. This paper investigates the view's implications for logic. An interpretation in terms of dialectical dispositions is proposed for the proof rules of a bilateral sequent system. Rules that are sound with respect to this interpretation can be seen as generating an *intrinsic logic* of dialectical disposition expressivism. It is argued that such a logic will be very weak—weaker than the intersection of minimal logic and FDE.

Keywords: connectives, bilateralism, expressivism, pragmatism, logical consequence

1 Introduction

It is a familiar idea that the behavior of logical operators can be illuminated in terms of their use in circumstances of dialectical engagement—circumstances in which speakers challenge and meet challenges to each other's assertions. The best-known elaboration of this idea is the game-theoretic tradition of "dialogical logic" (Lorenzen & Lorenz, 1978); it has also been pursued within other approaches, such as inferentialism (Brandom, 2008; Lance, 2001) and expressivist pragmatism (Price, 1990, 1994, 2009). In each case, specifying the dialectical roles of operators has served as a way of providing a semantic grounding for a logical consequence relation. Thus Lorenzen aims to underwrite intuitionistic logic, Lance and Brandom arrive (respectively) at a weak relevance logic and classical logic, and Price claims that his account of negation's function supports classical logic. Questions have accordingly been raised about whether our discursive practices satisfy the assumptions that would be needed for deriving the claimed conclusions about logical consequence (Dutilh Novaes, 2021; Hodges, 2001; Marion, 2009).

---

[1] I would like to thank audiences at Logica 2023 and the University of Connecticut for helpful discussion. I am particularly grateful for comments by Julian Schlöder and Ryan Simonelli, and for suggestions by a reviewer.

Lionel Shapiro

The present paper concerns a less ambitious version of the dialectical approach, *dialectical disposition expressivism* (Shapiro, 2023, §5). This is a proposal about the functions of the types of logical complexity regimented in logician's English using sentential connectives 'and', 'or', and 'not'. The proposal, which will be presented in Section 2, is that these connectives let speakers convey certain dispositions with respect to dialectical engagement. Elsewhere (Shapiro, 2023, §4), I have argued that expressivist theorists have no reason to demand that their accounts of the functions of connectives should settle what is a logical consequence of what. As will be explained in Section 6, this is an upshot of *deflationism about logical consequence* (Shapiro, 2022). On that view, we should no more expect an expressivist account of logical vocabulary to settle questions about what follows logically from what than we would expect an expressivist account of moral vocabulary to settle questions about what is morally permissible.

Still, there remains a legitimate question: What logic, if any, *can* in fact be underwritten by this account of the connectives? We may understand the question as follows. Is there any formal consequence relation such that dialectical disposition expressivism endows that relation with something like the pragmatic significance logical consequence is thought to have?[2] If so, such a relation may be deemed an "intrinsic logic" of dialectical disposition expressivism.[3] Importantly, an advocate of dialectical disposition expressism can recognize such an intrinsic logic without identifying it as the relation of logical consequence.[4]

My aim is to assess whether dialectical disposition expressivism has an intrinsic logic. To this end, I exploit a parallel between clauses specifying the connectives' expressive functions and "bilateral" proof rules (Rumfitt, 2000; Smiley, 1996). Section 3 shows how such rules, when cast in sequent format, can be naturally interpreted in terms of dialectical dispositions. I then investigate which bilateral rules can be justified on this interpretation. Section 4 considers rules that involve connectives, while Section 5 considers "coordination rules," rules that (on the proposed interpretation) encapsulate principles concerning challenging and meeting challenges. My conclusion

---

[2] This question is raised, but not answered, in (Shapiro, 2023, p. 250 n. 29). I thank Julian Schlöder for pressing me to pursue it further.

[3] The phrase "intrinsic logic" is used in a loosely related sense by Brandom (2008, p. 139).

[4] Lance (2001, pp. 448–49) too suggests that some claims about what is a logical consequence of what may not "follow from considerations of the structure of...linguistic practice per se." However, in contrast to deflationism about consequence, he says that any such claim would be a "substantive epistemic claim."

Dialectical Dispositions and Logic

will be that the proof system consisting of justifiable bilateral rules yields a consequence relation weaker than the intersection of minimal logic and the paraconsistent and paracomplete logic FDE.

2 Expressive functions

Dialectical disposition expressivism holds that logical connectives serve to express dispositions with respect to moves in a "game of giving and asking for reasons" (Brandom, 1994).[5] These moves include both *asserting* and *rejecting* propositions. Asserting a proposition will be understood as assuming the responsibility to meet challenges, and thereby authorizing others to defer to one's assertion in meeting challenges to their own assertions (Brandom, 1994, pp. 171–72). Rejecting a proposition will be understood as expressing one's disposition to challenge assertions of the proposition.

I will presuppose a conception of *challenging* where a challenge stands in a strong tension with the challenged assertion. To start with, challenging an assertion places the asserter under an obligation to defend the assertion, by adducing warrant or neutralizing the challenge, on pain of having to withdraw their assertion. But this needn't suffice to constitute a challenge. For example, merely telling someone that there is *insufficient evidence* for their assertion won't count as a challenge. Here is how I propose to understand the difference. If one asserts a proposition for which one also claims there is insufficient evidence, this needn't undermine the authority of one's pronouncement about the lack of evidence. By contrast, when one asserts a proposition that one also challenges, this undermines the authority of the challenge as well as of the assertion.[6]

Simplifying for present purposes, I will speak of asserting/rejecting sentences rather than propositions, and consider only sentences of a language whose logical complexity is exhausted by the three sentential connectives $\wedge$, $\vee$, and $\neg$. Concerning conjunction, the hypothesis is that one who asserts $A \wedge B$ expresses a disposition she has with regard to that very assertion. Specifically, she expresses her being disposed thus:

---

[5]This section abridges, with some additions, the proposal in (Shapiro, 2023, §5).

[6]There is an affinity here with Brandom's characterization of *incompatible* propositions, on which commitment to both propositions precludes entitlement to *either* proposition (Brandom, 1994, p. 169). Unlike Brandom, however, I won't invoke a relation of incompatibility between propositions, or the notion of commitment to a proposition.

($\wedge$-c_i) She is prepared to recognize an interlocutor's rejection of A (likewise of B) as a challenge to her assertion.

($\wedge$-m_s) When an interlocutor has challenged her assertion, she is prepared to adduce, as a way to meet the challenge, any pair of available assertions of A and B.

Let me explain the terminology. In the labels for the clauses, 'c' and 'm' stand for *challenging* and a *meeting* a challenge, while subscripts 'i' and 's' specify whether this is being done by *interlocutor* or *speaker*. An (actual or potential) assertion counts as "available" to a speaker if it is either an assertion she is prepared to make, or an assertion by another speaker she is prepared to defer to, in meeting a challenge.

Importantly, the above clauses don't specify the *only* kinds of challenge an asserter of a conjunction will recognize, or the *only* way she will be prepared to meet challenges to her assertion. For example, one who asserts $A \wedge B$ will be prepared to recognize an interlocutor's rejection of $A \wedge B$ as a challenge, yet the function of $\wedge$ isn't explained in terms of its conveying that disposition. One might wonder why both ($\wedge$-c_i) and ($\wedge$-m_s) are needed to explain the function of $\wedge$. The reason is that it's possible to have disposition ($\wedge$-c_i) without disposition ($\wedge$-m_s), and also vice versa. The former possibility is often exemplified by asserters of $(A \wedge B) \wedge C$. The latter possibility is often exemplifed by asserters of $(A \wedge B) \vee C$.

Making an assertion that expresses the above-specified disposition can facilitate dialectical engagement. Consider a circumstance in which one would be prepared to meet a challenge to one's assertion of C by asserting both A and B. With the resource of conjunction, dispute about one's defense of C can take the form of the assertion and rejection of the sentence $A \wedge B$. An interlocutor may be prepared to reject this conjunction without being prepared to reject either conjunct.

A parallel benefit is provided by a disjunction connective. In asserting $A \vee B$, a speaker expresses her being disposed thus:

($\vee$-c_i) She is is prepared to recognize an interlocutor's pair of rejections of A and of B as a challenge to her assertion.

($\vee$-m_s) When an interlocutor has challenged her assertion, she is prepared to adduce, as a way to meet the challenge, any available assertion of A (likewise of B).

Expressing this disposition, too, is useful in dialectical engagement. Consider a circumstance in which one could meet a challenge to one's assertion of C by an assertion of A (if available) as well as by an assertion of B (if available). Without asserting either A or B (perhaps one would reject A, while the interlocutor would reject B), one may meet the challenge by asserting their disjunction. With the resource of disjunction, dispute about one's defense of C can take the form of assertion and rejection of $A \vee B$.

Finally, following Price (1990), we can see negation as providing a means of rejecting a sentence by asserting another sentence. In asserting $\neg A$, a speaker expresses her being disposed thus:

($\neg$-c_s) She is prepared to challenge any assertion of A.

($\neg$-m_s) When her assertion is challenged, she is prepared to adduce, as a way to meet the challenge, any available assertion she would recognize as a way to challenge assertions of A.

Whereas the dispositions expressed by asserting a conjunction or disjunction all concern *that very assertion*, only one of the two dispositions expressed by asserting a negation does so, namely ($\neg$-m_s). A connective may also be such that asserting a sentence with it as major connective expresses dispositions that concern only the sentence's proper constituents. In (Shapiro, 2018), I propose that the dialectical dispositions expressed by a type of conditional concern only assertions and rejections of the *antecedent* and *consequent*.

The above expressive clauses help us understand the *raison d'être* of propositional logical complexity. Being able to assert logically complex sentences serves a purpose that would otherwise require including, as moves in the game of giving and asking for reasons, a hierarchy of distinct speech acts involving pluralities of sentences, starting with acts of asserting/rejecting pairs of sentences taken conjunctively or disjunctively. According to Humberstone (2000, p. 367–68), advocates of "bilateral" accounts of connectives in terms of assertion and rejection must explain why they don't invoke additional act of asserting conjunctively and disjunctively, acts standing to *conjunction* and *disjunction* the way rejection stands to *negation*. The present proposal yields a reply. The point of conjunction/disjunction is to let us do without asserting and rejecting conjunctively/disjunctively, whereas negation doesn't let us do without rejecting, which was invoked in explaining the functions of all three connectives.[7] Negation does, however, let us do without a distinct speech act of (say) disjunctively asserting one sentence together with another sentence taken negatively.

---

[7]Ripley (2020, p. 59 n. 7) likewise replies to Humberstone by noting the explanatory priority that bilateralism accords to rejection.

3 Interpreting sequent rules

In the cases of conjunction and disjunction, the above pairs of expressive clauses bear a suggestive resemblance to rules of a bilateral natural deduction system, specifically the negative and affirmative introduction rules. Here are these rules for conjunction:

$$\frac{-A}{-A \wedge B} \quad \frac{-B}{-A \wedge B} \; (-\wedge \mathrm{I}) \qquad \frac{+A \quad +B}{+A \wedge B} \; (+\wedge \mathrm{I})$$

Such rules are usually interpreted in terms of conditions on warranted or coherent assertion and rejection (Ripley, 2017). But their resemblance to our expressive clauses ($\wedge$-c_i) and ($\wedge$-m_s), respectively, motivates pursuing an alternative interpretation in terms of dialectical dispositions. For the purpose of using such an interpretation to build a consequence relation, it will be useful to reformulate the rules in sequent style. Here Γ is a set of signed sentences:

$$\frac{\Gamma \Rightarrow -A \, [-B]}{\Gamma \Rightarrow -A \wedge B} \; (-\wedge \mathrm{R}) \qquad \frac{\Gamma \Rightarrow +A \quad \Gamma \Rightarrow +B}{\Gamma \Rightarrow +A \wedge B} \; (+\wedge \mathrm{R})$$

To begin with, we define two relations between Γ and a signed sentence ϕ. Both relations will be relativized to an agent a, and for convenience both will be expressed in our metalanguage using the same ambiguous notation '$\Gamma \vdash_a \phi$'. The turnstile will receive different interpretations depending on whether ϕ carries positive or negative sign. In this respect, the current approach resembles the use of "dual" turnstiles for proof and refutation by Wansing (2017) and Ayhan (2021). The first relation, between Γ and negatively signed ϕ, corresponds to an agent's disposition with regard to *recognizing challenges* to their assertion.

Definition 1 $\Gamma \vdash_a -C$ iff a is disposed to recognize the following combination of speech acts by an interlocutor as challenging any assertion by a of C: the assertion of each positively signed member of Γ together with the rejection of each negatively signed member of Γ.

To see how this relation may apply, suppose that a is an "ideal discursive agent," one who (i) always *exhibits* all dialectical dispositions she expresses and (ii) knows which dialectical dispositions any assertion by her or by her interlocutors would express. I will now argue for the following conditional:

(1) If $\Gamma \vdash_a -A$ or $\Gamma \vdash_a -B$, then $\Gamma \vdash_a -A \wedge B$.

Suppose that a asserts $A \wedge B$. According to clause ($\wedge$-c_i), she thereby expresses a disposition to recognize an interlocutor's rejection of A as a challenge. As an ideal discursive agent, she will know that she in fact possesses this disposition. Now suppose, in addition, that there is a certain combination of assertions and rejections such that if a were to assert A, she would be prepared to recognize that combination as jointly challenging her assertion of A. Knowing this about herself, a will presumably be prepared to recognize an interlocutor who makes that combination of assertions and rejections as challenging her assertion of $A \wedge B$.

A second definable relation, this time between Γ and a positively signed ϕ, corresponds to a disposition with regard to adducing assertions to *meet challenges*. Here a bit of additional complexity is needed to accommodate an asymmetry between assertion and rejection. Whereas both assertions and rejections can count as challenging an assertion, such challenges can be met by further assertions, but not by rejections.

Definition 2 $\Gamma \vdash_a +C$ iff a is disposed to recognize the following combination of available assertions as meeting any challenge to any assertion by a of C: assertions of each positively signed member of Γ together with assertions, for each negatively signed member of Γ, of some sentence(s) that a regards as challenging that sentence's assertion.

An argument parallel to the one previously given uses clause ($\wedge$-m_s) to show, for any ideal agent a

(2) If $\Gamma \vdash_a +A$ and $\Gamma \vdash_a +B$, then $\Gamma \vdash_a +A \wedge B$.

In terms of the dual relations just defined, we can specify a pragmatic interpretation of an arbitrary sequent rule with n premises.

$$\frac{\Gamma_1 \Rightarrow \phi_1 \quad \ldots \quad \Gamma_n \Rightarrow \phi_n}{\Delta \Rightarrow \psi} \text{ (R)}$$

Definition 3 Sequent rule R is *DDE-sound* iff for all its instances and any ideal discursive agent a, if $\Gamma_i \vdash_a \phi_i$ for all $i \leq n$, then $\Delta \vdash_a \psi$.

The reasoning sketched above for (1) and (2) supports the claim that the rules $-\wedge R$ and $+\wedge R$ are DDE-sound. Here I should call attention to the fact that the reasoning assumed, in effect, that the two relations written '$\Gamma \vdash_a \phi$' exhibit a transitivity in virtue of which the following rule is DDE-sound.

$$\frac{\Gamma \Rightarrow \phi \quad \Gamma, \phi \Rightarrow \psi}{\Gamma \Rightarrow \psi} \text{ (Cut)}$$

The claim that Cut is DDE-sound is hardly indisputable. However, in the interest of investigating how strong an intrinsic logic can be defended using plausible assumptions, I propose to recognize Cut as DDE-sound unless we find that doing so would stand in tension with claiming DDE-soundness for other rules for which that claim is no less plausible. (One such consideration will emerge in Section 4.3.)

We can now define the notion that will be our central interest:

Definition 4 Let $\mathcal{G}$ be a set of unsigned sentences, and Γ the set of their positively signed counterparts. Then the sentences in $\mathcal{G}$ have A as a *DDE-intrinsic consequence* iff the "all-positive" sequent $\Gamma \Rightarrow +A$ is derivable in a system of DDE-sound rules.

Our question, now, concerns what additional sequent rules are DDE-sound, and thus how strong an intrinsic logic dialectical disposition expressivism will yield via our pragmatic interpretation of sequent rules.

4 Connective rules and initial sequents

It makes sense to consider first the full set of standard bilateral connective rules for $\wedge$, $\vee$, and $\neg$. These rules in natural-deduction format are known to yield the Belnap-Dunn logic FDE (Tamminga & Tanaka, 1999).[8] I will instead use corresponding Gentzen-style rules. The following system G_0 consists of one structural rule giving us initial sequents, in addition to four rules for each connective: positively and negatively signed left and right introduction rules.

$$\frac{}{\Gamma, \phi \Rightarrow \phi} \text{ (Id)}$$

$$\frac{\Gamma, +A\ [+B] \Rightarrow \phi}{\Gamma, +A \wedge B \Rightarrow \phi} \text{ (+}\wedge\text{L)} \qquad \frac{\Gamma \Rightarrow +A \quad \Gamma \Rightarrow +B}{\Gamma \Rightarrow +A \wedge B} \text{ (+}\wedge\text{R)}$$

$$\frac{\Gamma, -A \Rightarrow \phi \quad \Gamma, -B \Rightarrow \phi}{\Gamma, -A \wedge B \Rightarrow \phi} \text{ (-}\wedge\text{L)} \qquad \frac{\Gamma \Rightarrow -A\ [-B]}{\Gamma \Rightarrow -A \wedge B} \text{ (-}\wedge\text{R)}$$

---

[8]They are the connective rules in (Rumfitt, 2000, pp. 800–802). Rumfitt is ultimately interested in a system for classical logic. But, as Gibbard (2002) observes, Rumfitt's connective rules, in the absence additional "coordination principles" to be considered later, yield a "non-classical *constructive logic with strong negation*," namely Nelson's N4, of which FDE is the nonimplicative fragment (Omori & Wansing, 2017).

$$\frac{\Gamma, +A \Rightarrow \phi \quad \Gamma, +B \Rightarrow \phi}{\Gamma, +A \vee B \Rightarrow \phi} \; (+\vee L) \qquad \frac{\Gamma \Rightarrow +A \; [+B]}{\Gamma \Rightarrow +A \vee B} \; (+\vee R)$$

$$\frac{\Gamma, -A \; [-B] \Rightarrow \phi}{\Gamma, -A \vee B \Rightarrow \phi} \; (-\vee L) \qquad \frac{\Gamma \Rightarrow -A \quad \Gamma \Rightarrow -B}{\Gamma \Rightarrow -A \vee B} \; (-\vee R)$$

$$\frac{\Gamma, -A \Rightarrow \phi}{\Gamma, +\neg A \Rightarrow \phi} \; (+\neg L) \qquad \frac{\Gamma \Rightarrow -A}{\Gamma \Rightarrow +\neg A} \; (+\neg R)$$

$$\frac{\Gamma, +A \Rightarrow \phi}{\Gamma, -\neg A \Rightarrow \phi} \; (-\neg L) \qquad \frac{\Gamma \Rightarrow +A}{\Gamma \Rightarrow -\neg A} \; (-\neg R)$$

Proposition 1 *G_0 derives all and only the all-positive sequents corresponding to the consequences of FDE, and the signed Cut rule is admissible.*

Proof sketch. Consider the translation from signed to unsigned sequents that removes positive signs and replaces negative signs by $\neg$ (Humberstone, 2000, p. 365). Let G_0' be the unsigned system consisting of the translations of the rules of G_0, except for the (trivially redundant) translations of $+\neg L$ and $+\neg R$. G_0' is a system for FDE, a close variant of the cut-free system $\mathbf{LE}_{\text{fde2}}$ in (Anderson & Belnap, 1975, p. 179).[9]

We show that an all-positive sequent is derivable in G_0 iff its translation is derivable in G_0'. Left to right: any G_0-derivation translates into a G_0'-derivation. Right to left: take any G_0'-derivation of the unsigned sequent that translates the given all-positive sequent. For each step, consider the inference resulting from affixing positive signs to all sentences. It suffices to show that this inference is admissible in G_0. The only non-trivial cases are instances of *negated connective rules* of G_0'. Here we use the fact (shown by induction on derivation height) that $+\neg L$ and $+\neg R$ of G_0 are invertible. For example, consider the rule of G_0' that derives $\Gamma \Rightarrow \neg(A \wedge B)$ from $\Gamma \Rightarrow \neg A$. Let Γ^+ be the result of affixing positive signs to all members of Γ. Since $+\neg R$ is invertible, the inference from $\Gamma^+ \Rightarrow +\neg A$ to $\Gamma^+ \Rightarrow -A$ is admissible in G_0, and the latter sequent derives $\Gamma^+ \Rightarrow +\neg(A \wedge B)$. The admissibility of Cut in G_0 likewise follows from the admissibility of unsigned Cut in G_0'. □

---

[9]$\mathbf{LE}_{\text{fde2}}$ is identical to G_B in (Pynko, 1995). The differences between it and G_0' are just that $\mathbf{LE}_{\text{fde2}}$ has sequences as antecedents and succedents, and that the unsigned rules corresponding to $+\wedge L, -\wedge R, +\vee R$ and $-\vee L$ take "multiplicative" form. It is straightforward to show that if Σ is a sequence containing all and only members of the set Γ, then $\Gamma \Rightarrow \phi$ is derivable in G_0' iff $\Sigma \Rightarrow \phi$ is derivable in $\mathbf{LE}_{\text{fde2}}$.

Lionel Shapiro

4.1 A starting point

Which rules of G_0 are DDE-sound? Six rules are directly underwritten by our expressive clauses. We have already seen, in discussing claims (1) and (2), how the clauses for conjunction support $+\wedge$R and $-\wedge$R. Parallel reasoning uses expressive clauses ($\vee$-m_s) and ($\vee$-c_i) to justify the corresponding rules for disjunction, $+\vee$R and $-\vee$R. Additionally, in the case of negation, ($\neg$-c_s) will justify $+\neg$L, while ($\neg$-m_s) will justify $+\neg$R.

Admittedly, the rationale for including side formulas Γ in the initial sequents given by (Id) may not seem compelling. One counterexample might seem to be $-A, +A \vdash_a +A$. Will an asserter of A ever recognize, as meeting a challenge to her assertion, the puzzling combination of an assertion of A together with an assertion of something she takes to challenge A? I'll assume that if she won't do so, this will be because that pair of assertions won't count as available to her. Indeed, it isn't clear that they will ever be available. Even if she is prepared to assert both A and $\neg A$, perhaps in response to paradox, she may take both assertions' authority to be compromised.[10]

A more problematic case might seem to be $+A, -A \vdash_a -A$. Will an asserter of A be disposed to recognize an interlocutor who rejects A, while also asserting it, as having issued a challenge? I'll assume that if the asserter won't be so disposed, this is because she views the interlocutor's rejection of A as an act whose authority is undercut by his simultaneous assertion of A. However, it will be convenient not to modify Definition 1 to require that the interlocutor's challenge be recognized as one that carries authority, thereby avoiding $+A, -A \vdash_a -A$ while making the turnstile nonmonotonic. My goal is to argue that dialectical disposition expressivism generates at most a very weak consequence relation. As was the case with transitivity, this goal is served by not questioning the monotonicity of the relations defined in Definitions 1 and 2, unless we find that preserving monotonicity prevents us from recognizing additional rules as DDE-sound.[11]

---

[10] Priest (2006a) holds that while it's impossible to assert and reject the same sentence, we may sometimes be rationally required to. On the present view, doing so is possible, but only at the price of undercutting the authority of both acts—something we may find rationally unavoidable. However, the view is also compatible with holding (with Field, 2008) that we should reject a Liar sentence but *not* do so by asserting its negation.

[11] Still, the potential for nonmonotonicity shouldn't be exaggerated. For example, the mere fact that an agent a is disposed, *in some context*, to meet a challenge to her assertion of 'The match will light' by asserting 'The match was struck' doesn't make it the case that 'The match was struck' $\vdash_a$ 'The match will light'. Thus the failure of 'The match was struck', 'The match was wet' $\vdash_a$ 'The match will light' is no violation of monotonicity.

Dialectical Dispositions and Logic

4.2 Three other defensible rules

What can we say about the remaining six rules of G_0? A strong case can be made for the DDE-soundness of three of them: $+\wedge$L, $+\vee$L, and $-\neg$R. For the first two rules, a full discussion would require separate arguments for the cases where ϕ is positively and negatively signed. Here I will instead choose one case for each rule, as the relevant considerations extend to the case involving the other sign.

Starting with $+\wedge$L, take the case where ϕ is negatively signed. Our task is to justify the claim that if $\Gamma, A \vdash_a -C$, then $\Gamma, A \wedge B \vdash_a -C$. Suppose that an ideal discursive agent a would recognize an interlocutor's assertion of A as constituting part of a challenge to her assertion of C. Now, since she is an ideal discursive agent, a recognizes that someone who asserts $A \wedge B$ thereby overtly undertakes responsibility to meet any challenges to an assertion of A. Hence she will presumably take an interlocutor's status as having challenged her assertion of C to be preserved if, in place of asserting A, the interlocutor instead asserts $A \wedge B$.[12] Similar reasoning applies in the case of instances of $+\wedge$L where ϕ is positively signed.

For $+\vee$L, choose this time the case where ϕ is positively signed. Our task is to justify the claim that if $\Gamma, +A \vdash_a +C$ and $\Gamma, +B \vdash_a +C$, then $\Gamma, +A \vee B \vdash_a +C$. Suppose that an ideal discursive agent a takes it that a challenge to her assertion of C would be met (in the context of some other speech acts) by an available assertion of A, and that the challenge would likewise be met if that assertion were replaced by an available assertion of B. Now a will recognize that someone who asserts $A \vee B$ thereby overtly undertakes responsibility to meet any concurrent challenges to *both* A and B. But she recognizes that assertion of *either* of these sentences would suffice to meet a challenge to her assertion of C. Hence, she will presumably count an available assertion of $A \vee B$ as likewise meeting the challenge to C.[13] Again, similar reasoning applies in the case of instances of $+\vee$L where ϕ is negatively signed.

Finally, consider $-\neg$R and the claim that if $\Gamma \vdash_a +A$, then $\Gamma \vdash_a -\neg A$. Since part of what an assertion of $\neg A$ expresses is the speaker's disposition

---

[12]Here I assume the plausibility of the following principle. When someone has, in asserting, overtly committed themselves to meeting challenges to A, their assertion would count as challenging any assertions that would be challenged by asserting A.

[13]Here I assume the plausibility of the following principle. When someone has, in asserting, overtly committed themselves to meeting concurrent challenges to A and B, then their assertion (if available) meets challenges that could be met by available assertions of A as well as by available assertions of B.

Lionel Shapiro

to challenge assertions of A, it's reasonable to expect that one who asserts $\neg A$ will in turn recognize an interlocutor's assertion of A as a challenge to her own assertion. In reaching this conclusion, I'm not assuming that speakers regard the relation of challenging as symmetric—indeed, I'll offer a counterexample to that generalization in Section 5. Rather, the pair A and $\neg A$ is a special case. That's because I'm not merely relying on the claim that an ideal discursive agent who asserts A will in fact recognize assertions of $\neg A$ as challenges. Rather, I'm using the stronger claim that ideal discursive agents will recognize the asserting of $\neg A$ as a *way to express* that one is prepared to challenge all assertions of A.

4.3 Three problematic rules

That leaves three rules of G_0 to consider, namely the negative left introduction rules $-\wedge L$, $-\vee L$, and $-\neg L$. Each of these rules derives a conclusion about how ideal speakers are disposed to regard rejections of logical compounds. For example, justifying the DDE-soundness of $-\vee L$ requires justifying the claim that if $\Gamma, -A \vdash_a \phi$, then $\Gamma, -A \vee B \vdash_a \phi$. But our account of the expressive role of disjunction doesn't appear to have any consequences for how speakers will regard their own, or an interlocutor's, *rejection* of a disjunction. The prospects of a direct pragmatic justification of these rules look dim.

There is a general point here: the expressive function attributed to a connective by dialectical disposition expressivism is served in *assertions* of sentences with that major connective, not in rejections. In that sense, the position isn't fully "bilateral" — in understanding the role of the disjunction connective, asserting disjunctions has explanatory priority over rejecting them, even though what is expressed by asserting disjunctions is in turn understood in terms of the *assertion and rejection* of their disjuncts.

If the three negative left introduction rules can't be justified as DDE-sound, how weak will this leave the consequence relation generated by the remaining ones? Let G_1 be given by all rules of G_0 except for $-\wedge L$, $-\vee L$, and $-\neg L$.

Proposition 2 *The consequence relation corresponding to the derivability of all-positive sequents in G_1 is weaker than the intersection of FDE and minimal logic.*

Proof sketch. Each of the rules in G_1, in addition to $-\vee L$, is derivable in the system for minimal logic obtained by omitting the initial sequent

Dialectical Dispositions and Logic

$+A, -A \Rightarrow \phi$ from Humberstone's signed system for intuitionistic logic (Humberstone, 2000, p. 365). On the other hand, that system derives all-positive sequents that are not derivable in G_1. For example, an easy induction shows that a sequent with $+\neg A$ in the antecedent is derivable in G_1 only when A is a subformula of the succedent. Hence G_1 fails to derive the triple-negation elimination sequent $+\neg\neg\neg A \Rightarrow +\neg A$. Furthermore, G_1 fails to derive the De Morgan sequent $+\neg(A \vee B) \Rightarrow +\neg A$. □

4.4 An additional connective rule?

So far, we have only been looking at connective rules based on standard bilateral proof systems. Might dialectical disposition expressivism motivate additional connective rules on the present interpretation? The obvious candidate would be a form of disjunctive syllogism:[14]

$$\frac{\Gamma, +B \Rightarrow \phi}{\Gamma, -A, +A \vee B \Rightarrow \phi} \; (+\vee L')$$

We might seek to justify this rule by hypothesizing that in asserting $A \vee B$, a speaker expresses her being disposed thus:

($\vee$-m_s') When an interlocutor has challenged her assertion of B, she is prepared to recognize, as a way to meet the challenge, any available assertion she would recognize as a way to challenge A.

Does $+\vee L'$ plausibly reflect the expressive role of 'or'? If so, and we maintain our supposition that Cut is DDE–sound, then $+\vee L'$ can't be DDE-sound. To see why, consider the derivation

$$\cfrac{\cfrac{+A, +B \Rightarrow +B}{+A, -A, +A \vee B \Rightarrow +B} \; (+\vee L') \quad \cfrac{-A, +A \Rightarrow +A}{-A, +A \Rightarrow +A \vee B} \; (+\vee R)}{+A, -A \Rightarrow +B} \; (\text{Cut})$$

Do we really wish to say that an ideal agent will be disposed to take a challenge to their assertion of any sentence B to be met by available assertions of A together with a sentence they take to challenge the assertion of A? Even

---

[14]See (Mares, 2004, pp. 184–85). The symmetric rule for conjunction discussed there would derive $\Gamma, +A, -A \wedge B \Rightarrow \phi$ from $\Gamma, -B \Rightarrow \phi$. This is an unpromising candidate for DDE-soundness, due to the asymmetry between assertion and rejection explained in Section 4.3.

granting that such a combination of assertions may never be available, the dispositional claim is dubious.[15]

It appears, then, that at least one step in the above derivation isn't justified by the dialectical dispositions of ideal discursive agents. In support of $+\vee L'$ rather than Cut as the culprit, clause ($\vee$-m_s') can be disputed. In certain contexts, asserting $A \vee B$ may not convey that a speaker is disposed to recognize available assertions that challenge A as meeting challenges to assertions of B. Such contexts include ones where it's presupposed that the speaker's only way to meet a challenge to $A \vee B$ would be to assert A.

5 Coordination rules

At this point, a natural thought is that we may need to take a different route to justifying a stronger intrinsic consequence relation. So far, we have only considered *connective rules* in addition to (Id). But standard bilateral systems also employ also *non-connective* rules that Rumfitt (2000, p. 804) calls *coordination principles*: they "co-ordinate the assignment of positive and negative signs to particular contents." Is it possible that we can justify as DDE–sound coordination principles that are inadmissible in G_1, and perhaps even inadmissible in systems for FDE and/or minimal logic?

Following Smiley (1996), Rumfitt himself focuses on a coordination rule he calls "Smileian Reductio." Here $*$ reverses a sentence's sign.

$$\frac{\Gamma, \phi \Rightarrow \psi \quad \Gamma, \phi \Rightarrow \psi *}{\Gamma \Rightarrow \phi *} \text{ (SRed)}$$

The system resulting from adding SRed to G_0 is sound and complete with respect to classical consequence.[16] We can also consider the restricted version where ϕ is positively signed, which belongs to the rules of Humberstone's system for intuitionistic logic. This rule derives $+A, -A \Rightarrow -C$, yet it's

---

[15]If we add $+\vee L'$, Cut is no longer admissible, since there is no Cut-free derivation of $+A, -A \Rightarrow +B$. Extending G_0 with Cut and $+\vee L'$ has the effect of adding $\Gamma, +A, -A \Rightarrow \phi$ as an initial sequent. That's because the latter yields a proof system for the logic K$_3$, one that renders Cut and $+\vee L'$ admissible. To see that extending G_0' with $\Gamma, A, \neg A \Rightarrow B$ yields a system for K$_3$, note how G_0' and the extension mirror, respectively, the tableau systems for FDE and K$_3$ in (Priest, 2008); cf. (Beall, 2011, pp. 333–35).

[16]For completeness, it suffices to show that G_0+SRed renders admissible the natural deduction rules which, together with SRed, yield classical logic (Rumfitt, 2000, p. 804). To show this for the elimination rules, one can use Cut, which is derivable using SRed, Id, and a weakening rule admissible in G_0+SRed (Humberstone 2000, p. 351). For soundness, one can check that each rule is sound respect to the semantics in (Smiley, 1996).

hard to see how it will be the case that $+A, -A \vdash_a -C$. Surely an agent needn't be disposed to recognize the joint assertion and rejection of the same sentence as a challenge to any other assertion they have made.

More likely candidates for a DDE-sound coordination rule would be restricted versions of "Smileian Reversal," whose unrestricted formulation is as follows:[17]

$$\frac{\Gamma, \phi \Rightarrow \psi}{\Gamma, \psi* \Rightarrow \phi*} \text{ (SRev)}$$

Of the four rules this formulation encompasses, only the first two are admissible in Humberstone's system for intuitionistic logic.

$$\frac{\Gamma, +A \Rightarrow +B}{\Gamma, -B \Rightarrow -A} \text{ (R1)} \quad \frac{\Gamma, +A \Rightarrow -B}{\Gamma, +B \Rightarrow -A} \text{ (R2)}$$

$$\frac{\Gamma, -A \Rightarrow +B}{\Gamma, -B \Rightarrow +A} \text{ (R3)} \quad \frac{\Gamma, -A \Rightarrow -B}{\Gamma, +B \Rightarrow +A} \text{ (R4)}$$

None of these rules is admissible in G_0. Consider however their restrictions to empty Γ, henceforth R1'–R4'.

Proposition 3 *Each of R1'–R4' is admissible in G_0, but each is inadmissible in G_1.*

Proof sketch. Each of R1'–R4' is admissible in G_0 just in case its unsigned translation is admissible in G_0'. And those contraposition rules are all admissible in systems for FDE (cf. Priest, 2008, p. 162). As for G_1, adding R1' would derive $-A \vee B \Rightarrow -A$, adding R2' would derive $+\neg(A \vee B) \Rightarrow -A$, adding R3' would derive $-\neg A \Rightarrow +A$, and adding R4' would derive $+\neg\neg A \Rightarrow +A$. □

Can we justify any of R1'–R4' as DDE-sound? Interpreted as here proposed, these rules link *challenging* with *meeting challenges*. I'll argue that the linkages fail, and that we can understand why by understanding how the room left open by their failure is exploited by certain vocabulary. Specifically, I propose that epistemic modal operators "It might be the case that A" ($\Diamond A$) and "It must be the case that A" ($\Box A$) yield counterexamples to each of the restricted reversal rules.

---

[17] In systems where SRed and weakening are admissible, SRev is admissible as well, as its conclusion is derivable using SRed from the weakened premise $\Gamma, \psi*, \phi \Rightarrow \psi$ together with $\Gamma, \psi*, \phi \Rightarrow \psi*$.

$$\frac{+A \Rightarrow +\Box A}{-\Box A \Rightarrow -A} \qquad \frac{+\neg A \Rightarrow -\Diamond A}{+\Diamond A \Rightarrow -\neg A}$$

$$\frac{-A \Rightarrow +\Box \neg A}{-\Box \neg A \Rightarrow +A} \qquad \frac{-A \Rightarrow -\Diamond A}{+\Diamond A \Rightarrow +A}$$

Consider first the counterexample to the DDE-soundness of R1′. Arguably, an ideal discursive agent will be disposed to recognize available assertions of A as meeting a challenge to her own assertion of $\Box A$. Yet she won't be disposed to recognize an interlocutor's rejection of $\Box A$ as challenging her own assertion of A.

The counterexample to the DDE-soundness of R2′ is used by Lennertz (2019) to argue that the relation two agents stand in when *one disagrees with the other* can be asymmetric. In the present context, it amounts to an asymmetry pertaining to the *act of challenging*.[18] An ideal agent who asserts "It might snow in August" will be disposed to recognize an interlocutor's assertion of "It won't snow in August" as a challenge she needs to meet on pain of withdrawing her assertion. By contrast, an ideal agent who asserts "It won't snow in August" needn't be disposed to recognize an interlocutor's rejoinder of "It might snow in August" as a challenge she needs to meet on pain of withdrawal. The interlocutor has expressed his unwillingness to concede that it won't snow in August. However, he needn't be regarded as having challenged the asserter of "It won't snow in August" to defend her assertion on pain of withdrawal.[19]

6 Conclusion

We have been asking whether the expressive functions of connectives proposed by dialectical disposition expressivism endow any relation between sets of sentences and sentences with consequence-like pragmatic significance. The strongest consequence relation we have found a way to defend as being

---

[18] Simonelli (2024) argues that if challenging is understood, following Brandom, as asserting an *incompatible content* (commitment to which precludes entitlement to the challenged assertion's content), challenging must be symmetric. Responding to Lennertz's example, he leaves open the possibility that epistemic modals lack the kind of content that figures in Brandomian incompatibility. The present account doesn't appeal to incompatibility between contents.

[19] Compare Incurvati and Schlöder (2019, pp. 754–55, 759) on the speech act of "weak assertion" of A, which they describe as one that "prevents [$\neg A$] from being added to the common ground." They argue that assertions of $\Diamond A$ license A's weak assertion. As challenging is understood here, weakly asserting A doesn't amount to challenging assertions of $\neg A$.

(in this sense) *intrinsic* to dialectical disposition expressivism is generated by proof system G_1. This is a very weak consequence relation. It fails to validate any arguments invalid in minimal logic, including ones that are valid in FDE such as $\neg(A \wedge B) \vdash \neg A \vee \neg B$ or $\neg\neg A \vdash A$; it likewise fails to validate any arguments invalid in FDE, including ones that are valid in minimal logic such as $A, \neg A \vdash \neg B$. Furthermore, it even fails to validate some arguments that are valid in both minimal logic and FDE, such as $\neg\neg\neg A \vdash \neg A$ and $\neg(A \vee B) \vdash \neg A$.

Is the apparent weakness of its intrinsic logic an objection to dialectical disposition expressivism? It would only be an objection if that view aimed to give an explanation of the functions of connectives that accounts for logical consequences. Elsewhere, I argue that this aim would be out of place (Shapiro, 2023). That's because according to the deflationism about logical consequence defended in (Shapiro, 2022), talk of consequence is fundamentally *not* metalinguistic talk about sentences and their relations. Rather, consequence talk serves to let us generalize over *logical conditionals*. These are sentences whose major connective can be expressed in English using locutions like 'that ... entails that ...' (Anderson & Belnap, 1975, p. 491) or 'if ... then logically ...' (Priest, 2006b, p.82).[20] For example, by saying that every sentence of the form 'It isn't the case that it isn't the case that p' has 'p' as a logical consequence, we achieve the effect of generalizing over an infinite class of logical conditionals:

> If it isn't the case that snow is not white, then logically snow is white.
> If it isn't the case that theft is not wrong, then logically theft is wrong.
> etc.

On this view, inquiry into what is a logical consequence of what is only superficially about relations between sentences; it is more fundamentally inquiry into matters formulated using a logical conditional rather than a consequence predicate.

Now just as expressivists about the function of 'wrong' shouldn't be expected to show that their account settles whether *theft is wrong*, expressivists about the function of 'not' shouldn't be expected to show that their account settles whether *if it isn't the case that theft is not wrong, then logically theft is wrong*. Settling the former question involves doing ethics; settling the latter involves doing logic. In neither case should we expect that the

---

[20]Here 'logically' is to be understood as part of the conditional connective, rather than as expressing a modal operator that is part of its consequent.

question can be settled by studying the functions of words. Hence, on the deflationist approach to logical consequence, expressivists about the function 'not' shouldn't be expected to show that their account of that function settles whether instances of double negation elimination are cases of logical consequence.

In short, even if dialectical disposition expressivism yields only a very weak intrinsic logic, the position is compatible with holding that *logical consequence* is far stronger.

References

Anderson, A., & Belnap, N. (1975). *Entailment: The Logic of Relevance and Necessity* (Vol. 1). Princeton: Princeton University Press.

Ayhan, S. (2021). Uniqueness of logical connectives in a bilateralist setting. In M. Blicha & I. Sedlár (Eds.), *Logica Yearbook 2020* (pp. 1–16). Rickmansworth: College Publications.

Beall, J. (2011). Multiple-conclusion LP and default classicality. *Review of Symbolic Logic*, *4*, 326–336.

Brandom, R. (1994). *Making It Explicit*. Cambridge, Massachusetts: Harvard University Press.

Brandom, R. (2008). *Between Saying and Doing*. Oxford: Oxford University Press.

Dutilh Novaes, C. (2021). *The Dialogical Roots of Deduction*. Cambridge: Cambridge University Press.

Field, H. (2008). *Saving Truth from Paradox*. Oxford: Oxford University Press.

Gibbard, A. (2002). Price and Rumfitt on rejective negation and classical logic. *Mind*, *111*, 297–303.

Hodges, W. (2001). Dialogue foundations: A sceptical look. *Aristotelian Society Supplementary Volume*, *75*, 17–32.

Humberstone, L. (2000). The revival of rejective negation. *Journal of Philosophical Logic*, *29*, 331–381.

Incurvati, L., & Schlöder, J. (2019). Weak assertion. *Philosophical Quarterly*, *69*, 741–770.

Lance, M. (2001). The structure of linguistic commitment III: Brandomian scorekeeping and oncompatibility. *Journal of Philosophical Logic*, *30*, 439–464.

Lennertz, B. (2019). Might-beliefs and asymmetric disagreement. *Synthese*, *196*, 4775–4805.

Lorenzen, P., & Lorenz, K. (1978). *Dialogische Logik*. Darmstadt: Wissenschaftliche Buchgesellschaft.

Mares, E. (2004). *Relevant Logic*. Cambridge: Cambridge University Press.

Marion, M. (2009). Why play logical games? In O. Majer, A. Pietarinen, & T. Tulenheimo (Eds.), *Games: Unifying Logic, Language, and Philosophy* (pp. 3–26). Dordrecht: Springer.

Omori, H., & Wansing, H. (2017). 40 years of FDE: An introductory overview. *Studia Logica*, *105*, 1021–1049.

Price, H. (1990). Why 'not'? *Mind*, *99*, 221–238.

Price, H. (1994). Semantic minimalism and the Frege point. In S. Tsohatzidis (Ed.), *Foundations of Speech Act Theory* (pp. 132–155). London: Routledge.

Price, H. (2009). *'Not' again*. Unpublished manuscript. Retrieved from https://philpapers.org/rec/PRINA.

Priest, G. (2006a). *Doubt Truth to be a Liar*. Oxford: Oxford University Press.

Priest, G. (2006b). *In Contradiction* (2nd ed.). Oxford: Oxford University Press.

Priest, G. (2008). *An Introduction to Non-Classical Logic* (2nd ed.). Cambridge: Cambridge University Press.

Pynko, A. (1995). Characterizing Belnap's logic via De Morgan's laws. *Mathematical Logic Quarterly*, *41*, 442–454.

Ripley, D. (2017). Bilateralism, coherence, warrant. In F. Moltmann & M. Textor (Eds.), *Act-Based Conceptions of Propositional Content* (pp. 307–324). New York: Oxford University Press.

Ripley, D. (2020). Denial. In V. Déprez & M. Espinal (Eds.), *The Oxford Handbook of Negation* (pp. 47–57). Oxford: Oxford University Press.

Rumfitt, I. (2000). Yes and 'no'. *Mind*, *109*, 781–823.

Shapiro, L. (2018). Logical expressivism and logical relations. In O. Beran, V. Kolman, & L. Koreň (Eds.), *From Rules to Meanings* (pp. 179–195). New York: Routledge.

Shapiro, L. (2022). What is logical deflationism? Two non-metalinguistic conceptions of logic. *Synthese*, *200*(31), 1–28.

Shapiro, L. (2023). Neopragmatism and logic: A deflationary proposal. In J. Gert (Ed.), *Neopragmatism: Interventions in First-Order Philosophy* (pp. 235–257). Oxford: Oxford University Press.

Simonelli, R. (2024). Why must incompatibility be symmetric? *The Philosophical Quarterly*, *74*(2), 658–682.
Smiley, T. (1996). Rejection. *Analysis*, *56*, 1-9.
Tamminga, A., & Tanaka, K. (1999). A natural deduction system for First Degree Entailment. *Notre Dame Journal of Formal Logic*, *40*, 258-272.
Wansing, H. (2017). A more general proof theory. *Journal of Applied Logic*, *25*, S25–S47.

Lionel Shapiro
University of Connecticut, Philosophy Department
USA
E-mail: `lionel.shapiro@uconn.edu`

Generalized Bilateral Harmony

RYAN SIMONELLI[1]

Abstract: I introduce a schematic notation for formulating bilateral natural deduction systems, and I use this notation to formulate three distinct bilateral natural deduction systems for classical logic. I then propose a new criterion for bilateral harmony that I argue is superior to the existing criteria proposed in the literature. Finally, I show, at the schematic level, that all three bilateral systems meet this criterion of bilateral harmony.

Keywords: bilateralism, natural deduction, harmony

1 Introduction

Classical natural deduction famously suffers from a lack of harmony between the introduction and elimination rules. In response to this issue, Rumfitt (2000) argues that if one wants to account for the meanings of the classical connectives in terms of the rules governing their use in a natural deduction system, one should opt for a *bilateral* system, in which formulas are positively or negatively signed, expressing affirmations or denials. Such systems straightforwardly resolve the lack of harmony between introduction and elimination rules. However, they give rise to the further concern of lack of harmony between the positive and negative rules in the system, and some authors, such as Gabbay Gabbay (2017), have argued that this is a serious problem for bilateralism. In recent literature on bilateralism, several different proposals for bilateral harmony have been put forth in response to this concern (Francez, 2014a, Kürbis, 2022, del Valle-Inclan & Schlöder, 2023, del Valle-Inclan, 2023, Kürbis, 2021). I contend here, however, that all such proposals are unsatisfactory, either failing to rule out disharmonious connectives, ruling out harmonious ones, or achieving extensional adequacy at the expense of being *ad hoc*. This paper introduces a novel condition for bilateral harmony that is sufficient, necessary, and conceptually well-motivated,

---

[1] Many thanks to Kevin Davey, Bob Brandom, Ulf Hlobil, Pedro del Valle-Inclan, Lionel Shapiro and two anonymous referees for comments.

Ryan Simonelli

framing it within a broader, generalized approach to bilateralism. This generalized approach treats bilateral systems with a notation that schematizes over the polarity of positive or negative signs, enabling us to abstract away from specific bilateral rules and consider instead general bilateral rule forms and their various proof-theoretic virtues or vices.

This paper is structured as follows. In Section Two, I introduce a schematic notation that enables a specification of a bilateral system for all of the classical connectives in terms of a single rule schema. This enables me to schematically specify three distinct bilateral natural deduction systems for classical logic. In Section Three, I introduce the notions of unilateral and bilateral harmony as a pair of constraints that the rules of any bilateral natural deduction system must meet. In Section Four, I criticize the three existing approaches to bilateral harmony in the literature. Finally, in Section Five, I propose a new criterion of bilateral harmony and prove, at the schematic level, that all three systems I've specified meet this new criterion of bilateral harmony whereas problematic connectives, such as the bilateral version of *tonk*, fail to meet it.

2 Three Schematic Systems

There are two key innovations of bilateral natural deduction systems for classical logic of the sort proposed by Smiley (1996) and Rumfitt (2000).[2] The first key innovation is the rules for negation. In Rumfitt's system, they are the following:

$$\frac{-\langle\varphi\rangle}{+\langle\neg\varphi\rangle} +_{\neg I} \qquad\qquad \frac{+\langle\neg\varphi\rangle}{-\langle\varphi\rangle} +_{\neg E}$$

---

[2]Here, I limit my attention to bilateral natural deduction systems of the sort proposed by Smiley (1996) and Rumfitt (2000), primarily in the context of classical logic, in which formulas are positively and negatively signed. The authors discussed here all develop this style of bilateralism. In the past several years, a different style of bilateral system has been developed, primarily in the context of intuitionistic logic, by Wansing (2016), Wansing (2017), Ayhan (2021), Drobyshevich (2019), Wansing and Ayhan (2023), and others which involves a signing of the *turnstile* (or the horizontal deduction line) to express verification or falsification. There are interesting and important questions to ask about the relationship between these two styles of bilateralism. However, addressing those questions is left for another paper.

Generalized Bilateral Harmony

$$\frac{+\langle\varphi\rangle}{-\langle\neg\varphi\rangle}\, \neg_I \qquad\qquad \frac{-\langle\neg\varphi\rangle}{+\langle\varphi\rangle}\, \neg_E$$

These rules jointly codify that denying a sentence has the same logical significance as asserting its negation. They are obviously harmonious, and they clearly define an involutive negation, as both double negation introduction and elimination are immediately derived through two applications of the I-rules or E-rules respectively.

The second key innovation of Smiley/Rumfitt-style bilateral natural deduction systems are the *coordination principles*, structural rules which "coordinate" the opposite stances of affirmation and denial, formally codifying the sense in which these stances really are opposites. The standard formulation of bilateralism for classical logic, owed to Rumfitt, contains the following two, which I call "Incoherence" and "Reductio":[3]

$$\frac{A \quad A^*}{\bot}\; \text{Incoherence} \qquad\qquad \begin{array}{c}\overline{A}\;u \\ \vdots \\ \dfrac{\bot}{A^*}\; \text{Reductio}^u\end{array}$$

Here, A is any signed sentence, and starring a signed sentence yields the oppositely signed sentence. Thus, Incoherence says that from some stance A and its opposite A^*, one can conclude incoherence, and Reductio says that if, given the assumption of some stance A, one can conclude incoherence, then one can discharge that assumption and conclude the opposite stance, A^*. With these coordination principles, the negation rules given above yield classical negation.

---

[3] Smiley's original formulation of bilateralism for classical logic involved just one principle, which Rumfitt calls "Smiliean Reductio." There are other ways of specifying the coordination principles for classical logic. For instance, del Valle-Inclan (2023) proposes "Bilateral Explosion" and "Bilateral Excluded Middle." However, most recent proponents of bilateralism for classical logic (e.g. Kürbis (2021), Hjortland (2014), del Valle-Inclan and Schlöder (2023), Incurvati and Schlöder (2023)) have followed Rumfitt in using these two principles, and I will do so here. Given the inter-derivability of different equivalent sets of coordination principles, the main proposal for bilateral harmony presented in this paper can be implemented in systems containing different coordination principles.

Ryan Simonelli

This paper brings a third key idea to bilateralism.[4] Bilateral notation enables us to think of the rules for all of the binary connectives of classical logic as instances of general rule schemas. This enables us to consider different sets of binary connective rules and their respective proof-theoretic virtues and vices at a higher level of generality than standard approaches, abstracting from the polarity of signs (whether they are positive or negative) and just considering the opposition between stances towards sentences. To do this, I deploy a notation that schematizes over signs, using variables such as $\boldsymbol{a}$ and $\boldsymbol{b}$ to indicate signs that may be either $+$ or $-$ along with a function $*$ that maps $+$ to $-$ and $-$ to $+$. So, for any signed formula of the form $\boldsymbol{a}\langle\varphi\rangle$, where $\boldsymbol{a} \in \{+, -\}$, if $\boldsymbol{a} = +$ then $\boldsymbol{a}^* = -$, and if $\boldsymbol{a} = -$ then $\boldsymbol{a}^* = +$ (and so $\boldsymbol{a}^{**} = \boldsymbol{a}$).[5] With this notation introduced, we can consider rule sets as a whole in terms of their general form and do our proof-theory at this higher level of generality.

There are three bilateral systems I will consider here. All are extensions of the bilateral system originally proposed by Smiley (1996). Smiley proposes a bilateral system with just rules for negation, conjunction and disjunction. The rules for conjunction and disjunction both take the following form:

$$\frac{\boldsymbol{a}\langle\varphi\rangle \quad \boldsymbol{b}\langle\psi\rangle}{\boldsymbol{c}\langle\varphi \circ \psi\rangle} \boldsymbol{c}_{\circ I} \qquad \frac{\boldsymbol{c}\langle\varphi \circ \psi\rangle}{\boldsymbol{a}\langle\varphi\rangle} \boldsymbol{c}_{\circ E_1} \qquad \frac{\boldsymbol{c}\langle\varphi \circ \psi\rangle}{\boldsymbol{b}\langle\psi\rangle} \boldsymbol{c}_{\circ E_2}$$

Understanding the horizontal line as expressing commitment (e.g. Brandom, 1994, Incurvati & Schlöder, 2023), the $\boldsymbol{c}_{\circ I}$ rule says that if one takes stance $\boldsymbol{a}$ to φ and stance $\boldsymbol{b}$ to ψ, then one is committed to taking stance $\boldsymbol{c}$ to $\varphi \circ \psi$. The $\boldsymbol{c}_{\circ E}$ rules says that if one takes stance $\boldsymbol{c}$ to $\varphi \circ \psi$, then one is committed to taking stance $\boldsymbol{a}$ to φ and one is also committed to taking stance $\boldsymbol{b}$ to ψ. Though Smiley provides rules only for conjunction and disjunction that take this form, the whole set of standard binary connectives, along with several not so standard ones (the Sheffer Stroke, Perice's arrow, and the dual of the conditional), can be given rules of this form:

---

[4] See also Simonelli (2024a) for a development of this idea. A similar idea has been developed in the context of signed tableaux systems by Smullyan (1968), which are themselves a kind of bilateral system. However, the version of the schematic approach adopted here is both more flexible and more conceptually transparent.

[5] So that the star is not ambiguous, we might now say that, where A is shorthand for a formula of the form $\boldsymbol{a}\langle\varphi\rangle$, A^* is shorthand for $\boldsymbol{a}^*\langle\varphi\rangle$.

Generalized Bilateral Harmony

$\wedge$: $a = +, b = +, c = +$ $\vee$: $a = -, b = -, c = -$
$|$: $a = +, b = +, c = -$ $\downarrow$: $a = -, b = -, c = +$
$\rightarrow$: $a = +, b = -, c = -$ $\succ$: $a = -, b = +, c = +$
$\prec$: $a = +, b = -, c = +$ $\leftarrow$: $a = -, b = +, c = -$

The distinction between the rules for conjunction and disjunction and all of the other rules is that the conjunction and disjunction rules are *bilaterally homogeneous*, each containing only one sign, positive or negative, whereas all of the other rules are *bilaterally mixed*, containing both positive and negative signs. In a unilateral context, such "mixed" form would require appeal to negation, and so would violate the criterion of separability among the connective rules. In a bilateral context, however, there is no reason not to allow bilaterally mixed rules, and so this enables us to put forward rules for all of the connectives in terms of a single rule schema. I'll call this system, where rules for all of the connectives are given by this schema, BNK0.

It is easy to show, as Smiley does for the conjunctive and disjunctive fragment of this system, that, given the coordination principles, BNK0 is a sound and complete system for classical propositional logic containing all of these connectives. Still, although, in a *truth-functional* sense BNK0 is complete, in a *proof-theoretic* sense, it is not complete, since, for each connective, it contains only rules for affirming or denying sentences containing that connective. A proof-theoretically complete bilateral system must include rules for *both* affirming and denying each connective. I now want to consider three systems that complete BNK0 with rules specifying the grounds for and consequences of taking the opposite stance, c^*, to $\varphi \circ \psi$.

The first system I'll consider here is based on Rumfitt's (2000) system. Rumfitt supplements Smiley's positive conjunction and negative disjunction rules with negative conjunction and positive disjunction rules of the following form:

$$\dfrac{a^*\langle\varphi\rangle}{c^*\langle\varphi \circ \psi\rangle} c^*{\circ}I_1 \qquad \dfrac{b^*\langle\psi\rangle}{c^*\langle\varphi \circ \psi\rangle} c^*{\circ}I_2 \qquad \dfrac{c^*\langle\varphi \circ \psi\rangle \quad \begin{array}{c}\overline{a^*\langle\varphi\rangle}^{\,u} \\ \vdots \\ A\end{array} \quad \begin{array}{c}\overline{b^*\langle\psi\rangle}^{\,u} \\ \vdots \\ A\end{array}}{A} c^*{\circ}E$$

Now, in the system proposed by Rumfitt, the rules for the conditional are different in form from the rules for conjunction and disjunction. In the context of bilateral classical logic, however, there is no reason for this difference.

One way to see this is to see that the negative conditional rules proposed by Rumfitt are the following:

$$\frac{+\langle\varphi\rangle \quad -\langle\psi\rangle}{-\langle\varphi\to\psi\rangle}{-}{\to}\text{I} \qquad \frac{-\langle\varphi\to\psi\rangle}{+\langle\varphi\rangle}{-}{\to}\text{E}_1 \qquad \frac{-\langle\varphi\to\psi\rangle}{-\langle\psi\rangle}{-}{\to}\text{E}_2$$

These are of exactly the same form as the positive conjunction and negative disjunction rules he provides; they are of the form of the $c_\circ$ rules above. Insofar as this a uniform specification of the conditions and consequences of taking *one* stance (be it positive or negative) towards a conjunction, disjunction, or conditional, it's reasonable to think that the specification of the conditions and consequences of taking the *opposite* stance (be it negative or positive) towards a conjunction, disjunction, or conditional, should likewise be uniform. BNK1 provides such a uniform specification. It should clear, however, that it's not the only one.

Rather than providing rules that take the form of conjunction and disjunction in Rumfitt's system, we could alternatively provide a system in which the rules for all of the connectives have the form of Rumfitt's rules for the conditional. del Valle-Inclan and Schlöder (2023) have recently proposed such a system. In del Valle-Inclan and Schlöder's system, the $c_\circ$ rules are, once again, those of BNK0. So, the system contains the same same positive conjunction, negative disjunction, and negative conditional rules. However, the c^* rules are the following:

$$\frac{\overline{a\langle\varphi\rangle}^{\,u}}{\vdots} \\ \frac{b^*\langle\psi\rangle}{c^*\langle\varphi\circ\psi\rangle}c^*\circ_\text{I}{}^u \qquad\qquad \frac{c^*\langle\varphi\circ\psi\rangle \quad a\langle\varphi\rangle}{b^*\langle\psi\rangle}c^*\circ_\text{E}$$

Following Rumfitt's rules for the conditional, del Valle-Inclan and Schlöder treat the introduction rule with the hypothetical proof from $b\langle\psi\rangle$ to $a^*\langle\varphi\rangle$ and the elimination rule concluding $a^*\langle\varphi\rangle$ from $c^*\langle\varphi\circ\psi\rangle$ and $b\langle\psi\rangle$ as derived.[6] One can alternately treat these additional rules as basic. Either way, I'll call this second system, BNK2.

---

[6]The derivations go as follows:

Generalized Bilateral Harmony

Finally, I want to introduce a third bilateral system, with rules of a form that have not been previously considered, at least in a bilateral context.[7] Once again $c_\circ$ rules are kept as is, but we now use the following $c^*_\circ$ rules:

$$\cfrac{\overline{a\langle\varphi\rangle}^{\,u} \quad \overline{b\langle\psi\rangle}^{\,v} \\ \vdots \qquad \vdots \\ \bot}{c^*\langle\varphi\circ\psi\rangle}\,c^*{}_{\circ I}^{u,v} \qquad \cfrac{c^*\langle\varphi\circ\psi\rangle \quad a\langle\varphi\rangle}{b^*\langle\psi\rangle}\,c^*{}_{\circ E_1} \qquad \cfrac{c^*\langle\varphi\circ\psi\rangle \quad b\langle\psi\rangle}{a^*\langle\varphi\rangle}\,c^*{}_{\circ E_2}$$

Thus, we have the same elimination rules as BNK2 (I'll treat the second as primitive here), but a distinct introduction rule. I'll call this third system BNK3. The positive and negative introduction rules of BNK3, specified at this schematic level, fit together very intuitively. Once again, the $c_{\circ I}$ rule says that one is committed to taking stance c to $\varphi \circ \psi$ if one takes stance a to φ and stance b to ψ. The $c^*{}_{\circ I}$ rule, on the other hand, says that one is committed to taking the opposite stance, c^*, to $\varphi \circ \psi$ if taking a to φ along with taking stance b to ψ is incoherent.

So, to sum up, BNK0 provides rules for taking one stance, c, to each of the connectives. Whether this stance is positive or negative for any given connective is determined by the assignment of signs for connectives given above. All of the rules for taking stance c towards $\varphi \circ \psi$ have the same basic form as the familiar positive conjunction rules. The three systems considered here each supplement these $c_\circ$ rules with a set of rules for taking the opposite stance, c^*, towards $\varphi \circ \psi$. In particular, I've specified the following three systems:

1. **BNK1:** BNK0's $c_\circ$ rules + BNK1's $c^*_\circ$ rules (which all have the form of the familiar positive disjunction rules)

2. **BNK2:** BNK0's $c_\circ$ rules + BNK2's $c^*_\circ$ rules (which all have the form of the familiar positive conditional rules)

$$\cfrac{\cfrac{a\langle\varphi\rangle^2 \quad \cfrac{\overline{b\langle\psi\rangle}^{\,1}}{\mathcal{D}} \\ a^*\langle\varphi\rangle}{\cfrac{\bot}{b^*\langle\psi\rangle}\,\text{Red.}^1}\,\text{Inc.}}{c\langle\varphi\circ\psi\rangle}\,c^*{}_{\circ I}^{\,2} \qquad \cfrac{\cfrac{c\langle\varphi\circ\psi\rangle \quad \overline{a\langle\varphi\rangle}^{\,1}}{b^*\langle\psi\rangle}\,c^*{}_{\circ E} \quad b^*\langle\varphi\rangle}{\cfrac{\bot}{a^*\langle\varphi\rangle}\,\text{Red.}^1}\,\text{Inc.}$$

---
[7]Murzi (2020) considers such rules in a unilateral context, where they suffer from a problem of separability. There is no such problem in a bilateral context.

3. **BNK3:** BNK0's $c_\circ$ rules + BNK3's $c^*_\circ$ rules (which all have the new form shown above)

3 Unilateral and Bilateral Harmony

Having laid out these three systems, I now turn to the issue of the *harmony* among their rules. Let us start by first considering unilateral harmony, between introduction and elimination rules. Unilateral harmony means that the introduction and elimination rules are neither too strong nor too weak relative to each other. The classic case of disharmonious rules are those for the connective *tonk*, proposed first by Prior (1967), which has the following (positive) introduction and elimination rules:

$$\frac{+\langle \varphi \rangle}{+\langle \varphi \; tonk \; \psi \rangle} +tonk \; \mathrm{I}_1 \qquad \frac{+\langle \psi \rangle}{+\langle \varphi \; tonk \; \psi \rangle} +tonk \; \mathrm{I}_2$$

$$\frac{+\langle \varphi \; tonk \; \psi \rangle}{+\langle \varphi \rangle} +tonk \; \mathrm{E}_1 \qquad \frac{+\langle \varphi \; tonk \; \psi \rangle}{+\langle \psi \rangle} +tonk \; \mathrm{E}_2$$

The problem with *tonk* is that the elimination rules are too strong relative to the introduction rules. As such, it trivializes the logic, enabling us to conclude $+\langle \psi \rangle$ from $+\langle \varphi \rangle$ for arbitrary φ and ψ. A set of rules with the opposite problem are those for the connective that Francez (2015) calls *tunk*:

$$\frac{+\langle \varphi \rangle \quad +\langle \psi \rangle}{+\langle \varphi \; tunk \; \psi \rangle} +tunk \; \mathrm{I} \qquad \frac{+\langle \varphi \; tunk \; \psi \rangle \quad \overline{+\langle \varphi \rangle}^{\,u} \quad \overline{+\langle \psi \rangle}^{\,v}}{A} \;\; \overline{A} \;\; \overline{A}}{A} +tunk_E^{\,u,v}$$

Here, the elimination rule is *too weak* relative to the introduction rule. Though introducing a connective with these rules does not trivialize the consequence relation like introducing *tonk*, these rules are nevertheless disharmonious in an obvious sense, and a criterion for harmony ought to rule them out. So, any criterion of unilateral harmony ought to rule out *tonk* and *tunk* and do so in a systematic way.

A now standard approach to unilateral harmony, formulated by Pfenning and Davies (2001) expanding on a key idea of Prawitz (1965), is to conceive of harmony as established by a *reduction*, showing that the only consequences you can derive from a complex formula are among the grounds you used to derive it, and an *expansion* showing that, by extracting consequences from a complex formula, you can always recover the grounds required to derive it. The reduction shows that the elimination rules are not *too strong* relative to the introduction rules, whereas the expansion shows that the elimination rules are not *too weak* relative to the introduction rules. Schematically, these reductions and expansions for our $c_\circ$ rules, establishing unilateral harmony, go as follows:

$$\cfrac{\cfrac{\mathcal{D}_1 \quad \mathcal{D}_2}{\cfrac{\boldsymbol{a}\langle\varphi\rangle \quad \boldsymbol{b}\langle\psi\rangle}{\boldsymbol{c}\langle\varphi\circ\psi\rangle}\, c_{\circ\mathrm{I}}}}{\boldsymbol{a}\langle\varphi\rangle}\, c_{\circ\mathrm{E}_1} \qquad \cfrac{\cfrac{\mathcal{D}_1 \quad \mathcal{D}_2}{\cfrac{\boldsymbol{a}\langle\varphi\rangle \quad \boldsymbol{b}\langle\psi\rangle}{\boldsymbol{c}\langle\varphi\circ\psi\rangle}\, c_{\circ\mathrm{I}}}}{\boldsymbol{b}\langle\psi\rangle}\, c_{\circ\mathrm{E}_2} \quad \rightsquigarrow_r \quad \mathcal{D}_1 \quad \mathcal{D}_2$$
$$\boldsymbol{a}\langle\varphi\rangle \qquad \boldsymbol{b}\langle\psi\rangle$$

$$\begin{array}{c}\mathcal{D}_1 \\ \boldsymbol{c}\langle\varphi\circ\psi\rangle\end{array} \quad \rightsquigarrow_e \quad \cfrac{\cfrac{\mathcal{D}_1}{\boldsymbol{c}\langle\varphi\circ\psi\rangle}\, c_{\circ\mathrm{E}_1}\quad \cfrac{\mathcal{D}_1}{\boldsymbol{c}\langle\varphi\circ\psi\rangle}\, c_{\circ\mathrm{E}_2}}{\boldsymbol{c}\langle\varphi\circ\psi\rangle}\, c_{\circ\mathrm{I}}$$

This criterion of unilateral harmony enables us to rule out *tonk* and *tunk* and do so in a systematic way that reveals the sense in which they have the opposite problem: for *tonk*, no reduction is possible, whereas, for *tunk*, no expansion is possible.

The c^* rules of BNK1, BNK2, and BNK3 are all unilaterally harmonious as well. The forms of the c^* rules of BNK1 and BNK2 are familiar as those of the standard unilateral disjunction rules and conditional rules respectively, and the well-known reductions and expansions for those rules can be schematized to yield unilateral harmony proofs for BNK1 and BNK2. For BNK3, the reductions and expansions go as follows:

$$\cfrac{\overline{\boldsymbol{a}\langle\varphi\rangle}^{\,1} \quad \overline{\boldsymbol{b}\langle\psi\rangle}^{\,2}}{\cfrac{\cfrac{\mathcal{D}_1}{\bot}}{\boldsymbol{c}^*\langle\varphi\circ\psi\rangle}\, c^*_{\circ\mathrm{I}}{}^{1,2} \quad \cfrac{\mathcal{D}_3}{\boldsymbol{a}\langle\varphi\rangle}}{\boldsymbol{b}^*\langle\psi\rangle}\, c^*_{\circ\mathrm{E}_1} \quad \rightsquigarrow_r \quad \cfrac{\mathcal{D}_3 \quad \overline{\boldsymbol{b}\langle\psi\rangle}^{\,1}}{\cfrac{\cfrac{\boldsymbol{a}\langle\varphi\rangle}{\mathcal{D}_1}}{\bot}}{\boldsymbol{b}^*\langle\psi\rangle}\, \text{Reductio}^{\,1}$$

131

$$\begin{array}{c} \mathcal{D}_1 \\ \boldsymbol{c}^* \langle \varphi \circ \psi \rangle \end{array} \quad \rightsquigarrow_e \quad \dfrac{\dfrac{\begin{array}{c}\mathcal{D}_1\\ \boldsymbol{c}^*\langle \varphi \circ \psi\rangle\end{array} \quad \overline{\boldsymbol{a}\langle \varphi\rangle}^{\,1}}{\boldsymbol{b}^*\langle \psi\rangle}\boldsymbol{c}^*{\circ\mathrm{E}_1} \quad \overline{\boldsymbol{b}\langle \psi\rangle}^{\,2}}{\dfrac{\bot}{\boldsymbol{c}^*\langle \varphi\circ\psi\rangle}\boldsymbol{c}^*{\circ\mathrm{I}}^{1,2}}$$

So, all three systems are unilaterally harmonious. However, as several authors have pointed out, in a bilateral context, unilateral harmony between introduction and elimination rules is not enough.

To see why a further criterion of bilateral harmony is needed to supplement a criterion of unilateral harmony, consider the rules for the connective that I'll call *bonk*, presented first in the form of BNK1 rules:

$$\dfrac{+\langle \varphi \rangle}{+\langle \varphi \text{ bonk } \psi \rangle} +\text{bonk } \mathrm{I}_1 \qquad \dfrac{+\langle \psi \rangle}{+\langle \varphi \text{ bonk } \psi \rangle} +\text{bonk } \mathrm{I}_2$$

$$\dfrac{+\langle \varphi \text{ bonk } \psi \rangle \quad \begin{array}{c}\overline{+\langle \varphi\rangle}^{\,u}\\ \vdots \\ A\end{array} \quad \begin{array}{c}\overline{+\langle \psi\rangle}^{\,v}\\ \vdots \\ A\end{array}}{A} +\text{bonk}_E{}^{u,v}$$

$$\dfrac{-\langle \varphi \rangle}{-\langle \varphi \text{ bonk } \psi \rangle} -\text{bonk } \mathrm{I}_1 \qquad \dfrac{-\langle \psi \rangle}{-\langle \varphi \text{ bonk } \psi \rangle} -\text{bonk } \mathrm{I}_2$$

$$\dfrac{-\langle \varphi \text{ bonk } \psi \rangle \quad \begin{array}{c}\overline{-\langle \varphi\rangle}^{\,u}\\ \vdots \\ A\end{array} \quad \begin{array}{c}\overline{-\langle \psi\rangle}^{\,v}\\ \vdots \\ A\end{array}}{A} -\text{bonk}_E{}^{u,v}$$

These rules, which combine BNK1's positive disjunction rules with its negative conjunction rules, are unilaterally harmonious. Insofar as a harmony constraint is supposed to rule out *tonkish* connectives, there must be failure of bilateral harmony, for rules of this form, like those for *tonk*, enable us to conclude $+\langle q \rangle$ from $+\langle p \rangle$ for arbitrary atomics p and q:[8]

---

[8] In similar fashion, they also let us conclude $-\langle p \rangle$ from $+\langle q \rangle$, $-\langle q \rangle$ from $-\langle p \rangle$, and $+\langle p \rangle$ from $+\langle q \rangle$.

Generalized Bilateral Harmony

$$\dfrac{+\langle p\rangle}{+\langle p\ bonk\ q\rangle}\ +bonk\ I_1 \qquad \dfrac{\overline{-\langle q\rangle}^{\ 1}}{-\langle p\ bonk\ q\rangle}\ -bonk\ I_2$$

$$\dfrac{\bot}{+\langle q\rangle}\ \text{Red.}^1 \qquad \text{Inc.}$$

Note here that only the introductions rules are used to derive $+\langle q\rangle$ from $+\langle p\rangle$. The problem with *bonk*, and other connectives with rules of this same form, is that the grounds for concluding opposite stances towards $\varphi\ bonk\ \psi$, specified by the introduction rules, make it *too easy* to conclude opposite stances towards such a sentence, requiring *too little*. A criterion of bilateral harmony, then, must rule out these rules as disharmonious.

The opposite of *bonk* is a connective that Kürbis (2021) calls *conk*, which combines the positive conjunction rules with the negative disjunction rules:

$$\dfrac{+\langle\varphi\rangle\quad+\langle\psi\rangle}{+\langle\varphi\ conk\ \psi\rangle}\ +conk\ I \qquad \dfrac{+\langle\varphi\ conk\ \psi\rangle}{+\langle\varphi\rangle}\ +conk\ E_1 \qquad \dfrac{+\langle\varphi\ conk\ \psi\rangle}{+\langle\psi\rangle}\ +conk\ E_2$$

$$\dfrac{-\langle\varphi\rangle\quad -\langle\psi\rangle}{-\langle\varphi\ conk\ \psi\rangle}\ -conk\ I \qquad \dfrac{-\langle\varphi\ conk\ \psi\rangle}{-\langle\varphi\rangle}\ -conk\ E_1 \qquad \dfrac{-\langle\varphi\ conk\ \psi\rangle}{-\langle\psi\rangle}\ -conk\ E_2$$

Here, once again, we have unilateral harmony, but not bilateral harmony. In this case, however, grounds for introducing opposite stances to $\varphi\ conk\ \psi$, specified by the introduction rules, make it *too hard* to conclude opposite stances towards such a sentence, requiring *too much*. Accordingly, rather than using the introduction rules, we derive $+\langle q\rangle$ from $+\langle p\rangle$ using only the elimination rules:

$$\dfrac{\overline{-\langle p\ conk\ q\rangle}^{\ 1}}{-\langle p\rangle}\ -conk\ E_1 \qquad +\langle p\rangle \qquad \text{Inc.}$$

$$\dfrac{\bot}{+\langle p\ conk\ q\rangle}\qquad +conk\ E_2$$

$$\dfrac{}{+\langle q\rangle}$$

So, a criterion of bilateral harmony must rule out these rules as disharmonious as well. In the next section, I'll consider the three approaches that exist in the literature for ruling out connectives like *bonk* and *conk*.[9]

---

[9]There are a number of other bilaterally tonkish connectives that have been proposed in the literature (del Valle-Inclan and Schölder's *blink* and *bink*, Gabbay's (2017) •), but *bonk* and *conk* are representative examples of the two basic ways in which a connective can be bilaterally disharmonious.

Ryan Simonelli

4 Three Approaches to Bilateral Harmony

The first approach to bilateral harmony is owed to Francez (2014, 2015), and a variant of this approach has also been put forward by Kürbis (2022). Francez's criterion of bilateral harmony essentially works on the assumption, shared by Rumfitt (2000), that the starting point for a bilateral system ought to be a unilateral system of the sort proposed by Gentzen (1935). Given positive introduction rules of the form proposed by Gentzen (combining rules in the case of conjunction, splitting rules in the case of disjunction, and hypothetical rules, in the case of the conditional), Francez provides a recipe for determining corresponding negative rules, and his criterion of bilateral harmony is simply that the rules conform to this recipe. Francez's criterion rules out the rules for *conk* and *bonk*. For *conk*, since the positive introduction rule is a combining rule with two premises, the corresponding negative introduction rules must be two splitting rules, each one having, as its lone premise, the opposite of one of the premises of the positive introduction rule. For *bonk*, since the positive introduction rules are two splitting rules, the negative introduction rule must be a combining rule with two premises, each the opposite of the corresponding splitting rule.

Now, one basic conceptual problem with Francez's approach, noted by Kürbis (2021), is that the privileging of rules for affirmation, deriving rules for denial by "inversion," is antithetical to the basic philosophical commitment of bilateralism of treating affirmation and denial on a par with each other. However, even bracketing this conceptual problem, a more concrete problem is that it simply fails to provide an adequate general criterion of bilateral harmony. On the one hand, it simply rules out, without any justification, any of the rules of BNK2 where $c = +$ and all of the rules of BNK3. For a concrete case, consider BNK3's positive and negative introduction rules for conjunction:

$$\frac{+\langle \varphi \rangle \quad +\langle \psi \rangle}{+\langle \varphi \wedge \psi \rangle} +_{\wedge I} \qquad \frac{\overline{+\langle \varphi \rangle}^{\,u} \quad \overline{+\langle \psi \rangle}^{\,v}}{\frac{\vdots \qquad \vdots}{\frac{\bot}{-\langle \varphi \wedge \psi \rangle}}} -_{\wedge I}$$

These rules clearly *seem* harmonious. If one is going to claim that they are not harmonious, one ought to have a good argument for this claim. Francez's criterion of harmony simply rules them out by fiat. Even worse, consider positive and negative introduction rules for the connective *bonk*, in the form of the BNK3 rules:

$$\dfrac{\dfrac{}{-\langle\varphi\rangle}u \quad \dfrac{}{-\langle\psi\rangle}v}{\dfrac{\vdots \quad \vdots}{\dfrac{\bot}{+\langle\varphi \text{ bonk } \psi\rangle}}}+_{bonkI}{}^{u,v} \qquad \dfrac{\dfrac{}{+\langle\varphi\rangle}u \quad \dfrac{}{+\langle\psi\rangle}v}{\dfrac{\vdots \quad \vdots}{\dfrac{\bot}{-\langle\varphi \text{ bonk } \psi\rangle}}}-_{bonkI}{}^{u,v}$$

We can use these introduction rules and the coordination principles to derive $+\langle q\rangle$ from $+\langle p\rangle$ for arbitrary p and q:[10]

$$\dfrac{\dfrac{\dfrac{\overline{-\langle p\rangle}^{1} \quad +\langle p\rangle}{\bot} \text{Inc.}}{+\langle p \text{ bonk } q\rangle}+_{bonkI}{}^{1,0} \quad \dfrac{\dfrac{+\langle q\rangle \quad \overline{-\langle q\rangle}^{3}}{\bot}^{2}}{-\langle p \text{ bonk } q\rangle}-_{bonkI}{}^{2,0}}{\dfrac{\bot}{+\langle q\rangle}\text{Red.}^{3}} \text{Inc.}$$

Clearly, then, these rules are just as disharmonious as the above rules for *bonk*. Francez's criterion of bilateral harmony, however, is simply silent on whether these rules are harmonious or not, since he simply doesn't consider bilateral rules of this form. One might add some further criteria to cover rules of this form, but this seems hopelessly ad hoc, and won't ward against potential bilateral rules of yet different forms. A more general and principled approach is needed.

I now turn to a second approach to bilateral harmony, recently been proposed by del Valle-Inclan and Schlöder (2023). The approach begins by noting that, in the trivializing proofs above using *bonk* and *conk*, we use the coordination principles of Reductio and Incoherence on logically complex formulas. del Valle-Inclan and Schlöder's proposed bilateral harmony constraint, aimed at ruling out such problematic connectives, is to require that all coordination principles can be restricted to atomics. Notably, as Ferreira (2008) points out, BNK1 *fails* to meet this constraint; Reductio in particular cannot be restricted to atomics. To see this, consider the proof of the law of non-contradiction:

---

[10]Note here, we use vacuous discharges, as is standard in classical natural deduction systems.

$$\cfrac{\cfrac{\overline{+\langle p \wedge \neg p\rangle}^{\ 1}}{+\langle p\rangle}\ {+}_{\wedge\mathrm{E}_2}\quad \cfrac{\cfrac{\overline{+\langle p \wedge \neg p\rangle}^{\ 1}}{+\langle \neg p\rangle}\ {+}_{\wedge\mathrm{E}_2}}{-\langle p\rangle}\ {+}_{\neg\mathrm{E}}}{\cfrac{\bot}{-\langle p \wedge \neg p\rangle}\ \mathrm{Red.}^{\,1}}\ \mathrm{Inc.}$$

Here, we use Reductio on $+\langle p \wedge \neg p\rangle$, deriving an incoherence on the basis of this assumption and concluding the opposite, $-\langle p \wedge \neg p\rangle$. As Ferreira shows, there is no way to derive this formula without such a use of Reductio. While Ferreira concludes that this is a problem for bilateralism as such, del Valle-Inclan and Schlöder (2023) argue that this is just a problem for the specific rules that Rumfitt provides for conjunction and disjunction: rules of the form of BNK1. The BNK2 rules, as del Valle-Inclan and Schlöder show, meet this constraint. Schematizing the proofs they provide for this specific case where BNK1 fails, we have the following reduction:

$$\cfrac{\cfrac{\overline{\boldsymbol{c}\langle\varphi\circ\psi\rangle}^{\ 1}}{\mathcal{D}_1}}{\cfrac{\bot}{\boldsymbol{c}^*\langle\varphi\circ\psi\rangle}\ \mathrm{Red.}^{\,1}} \quad\rightsquigarrow\quad \cfrac{\cfrac{\overline{\boldsymbol{a}\langle\varphi\rangle}^{\ 1}\quad \overline{\boldsymbol{b}\langle\psi\rangle}^{\ 2}}{\boldsymbol{c}\langle\varphi\circ\psi\rangle}\ \boldsymbol{c}_{\circ\mathrm{I}}}{\cfrac{\mathcal{D}_1}{\cfrac{\bot}{\boldsymbol{b}^*\langle\psi\rangle}\ \mathrm{Red.}^{\,2}}\ \boldsymbol{c}^*_{\circ\mathrm{I}}{}^{1}}{\boldsymbol{c}^*\langle\varphi\circ\psi\rangle}}$$

Thus, in any case in which we assume $\boldsymbol{c}\langle\varphi\circ\psi\rangle$ and derive an incoherence to conclude $\boldsymbol{c}^*\langle\varphi\circ\psi\rangle$ by way of Reductio, we could just as well assume $\boldsymbol{a}\langle\varphi\rangle$ and $\boldsymbol{b}\langle\psi\rangle$, use Reductio only on $\boldsymbol{b}\langle\psi\rangle$, a formula of lesser complexity, and then conclude $\boldsymbol{c}^*\langle\varphi\circ\psi\rangle$ by way of the $\boldsymbol{c}^*$ introduction rule. BNK2, however, is not the only system that meets this constraint: BNK3 meets it just as well. For this specific case, we have the following reduction:

$$\cfrac{\cfrac{\overline{\boldsymbol{c}\langle\varphi\circ\psi\rangle}^{\ 1}}{\mathcal{D}_1}}{\cfrac{\bot}{\boldsymbol{c}^*\langle\varphi\circ\psi\rangle}\ \mathrm{Red.}^{\,1}} \quad\rightsquigarrow\quad \cfrac{\cfrac{\overline{\boldsymbol{a}\langle\varphi\rangle}^{\ 1}\quad \overline{\boldsymbol{b}\langle\psi\rangle}^{\ 2}}{\boldsymbol{c}\langle\varphi\circ\psi\rangle}\ \boldsymbol{c}_{\circ\mathrm{I}}}{\cfrac{\mathcal{D}_1}{\bot}}{\boldsymbol{c}^*\langle\varphi\circ\psi\rangle}}\ \boldsymbol{c}^*_{\circ\mathrm{I}}{}^{1,2}$$

Whereas, in the case of BNK2, we reduce the complexity of the Reductio'd formula, in the case of BNK3 shown here, we eliminate the use of Reductio entirely. Similar reductions can be given for all of the applications of coordination principles on logically complex formulas for BNK3.[11] So, on this second approach to bilateral harmony, BNK2 and BNK3 are bilaterally harmonious, whereas BNK1 is not.

This second approach to bilateral harmony fares much better than Francez's in providing a *general* constraint that can be applied to any set of rules, no matter their form. Unlike Francez's criterion, del Valle-Inclan and Schlöder's approach rules out the BNK3-form *bonk* rules. However, it still has the same basic problem: it can plausibly be regarded as a *sufficient* condition of bilateral harmony, but, conceived of as a *necessary* condition, it is too strong, ruling out intuitively harmonious rules as disharmonious. In this case, it is the BNK1 rules that get the boot. Once again, for a concrete case, consider just the positive and negative conjunction introduction rules of BNK1:

$$\frac{+\langle\varphi\rangle \quad +\langle\psi\rangle}{+\langle\varphi \wedge \psi\rangle} +_{\wedge\mathrm{I}} \qquad \frac{-\langle\varphi\rangle}{-\langle\varphi \wedge \psi\rangle} -_{\wedge\mathrm{I}_1} \qquad \frac{-\langle\psi\rangle}{-\langle\varphi \wedge \psi\rangle} -_{\wedge\mathrm{I}_2}$$

Once again, these positive and negative rules clearly *seem* to be harmonious. Indeed, it is just the intuition that "a conjunction [which is assertible upon the assertion of both conjunctions (together)] should be deniable upon the denial of each conjunct (separately)" that leads Francez (2015, 159) to propose his criterion of bilateral harmony which generalizes this intuition. Now, I've just argued that there may be harmonious rules of different forms that fail to meet Francez's specific criterion, and so it does not constitute a *necessary* criterion of bilateral harmony. Nevertheless, it still does seem to be a *sufficient* criterion in that positive and negative rules that meet it are neither too strong nor too weak relative to each other. At the very least, if one is going to claim that rules that meet it are disharmonious, one ought to have a good reason to do so. However, the only explicit motivation that del Valle-Inclan and Schlöder provide for imposing their criterion of harmony on a set of bilateral rules is that it rules out connectives like *bonk* and *conk*, but, as we've seen, requiring that rules conform to Francez's criterion of harmony suffices to rule out connectives like *bonk* and *conk* as well. So, once again, it seems that we are given a criterion of bilateral harmony that is *sufficient* but not *necessary*.

There is one more approach to bilateral harmony in the literature that I cannot go into here for reasons of space, and that is the approach articulated

---

[11] Since the elimination rules of BNK2 and BNK3 are the same, the reductions of Incoherence are the same. The reduction of the other direction of Reductio for BNK3, where $c^*\langle\varphi \circ \psi\rangle$ assumed to conclude $c\langle\varphi \circ \psi\rangle$ is similar to that for BNK2 shown by del Vall-Inclan and Schlöder.

by Kürbis (2021). Kürbis puts forward a normalization procedure for bilateral natural deduction systems with several reduction steps. While the various steps *do* yield an extensionally adequate criterion of bilateral harmony, there is no clear principle of unity among the various conditions, and so the approach itself ends up looking rather ad hoc. This is something that Kürbis himself admits, claiming that, while this approach technically gets the result and rules out connectives like *conk*, there is reason to think that it "does not really go to the heart of the matter of what is wrong with *conk*," (553). The criterion for unilateral harmony systematically rules out *tonk* and *tunk* in a way that is conceptually illuminating, getting to the heart of the matter as to what is wrong with these connectives. We should aspire to a criterion for bilateral harmony that does the same thing. Kürbis's account, self-admittedly, does not.

5 A New Approach to Bilateral Harmony

I now turn to the task of articulating a new criterion of bilateral harmony. It is not hard to articulate, informally, what is wrong with *bonk* and *conk*. As I've already said, intuitively, the problem with the positive and negative introduction rules for *bonk* is that it is *too easy* to introduce opposite stances towards φ *bonk* ψ. In particular, introducing opposite stances towards some sentence, which enables one to conclude an incoherence, should require grounds that are themselves incoherent. In the case of *bonk*, however, one can conclude $+\langle\varphi \text{ bonk } \psi\rangle$ and $-\langle\varphi \text{ bonk } \psi\rangle$, two opposite stances, from $+\langle\varphi\rangle$ and $-\langle\psi\rangle$, two stances which are not themselves guaranteed to be incoherent. On the other hand, once again very informally, the problem with the positive and negative introduction rules for *conk* is that it is *too hard* to introduce opposite stances towards φ *conk* ψ. In particular, to introduce $+\langle\varphi \text{ conk } \psi\rangle$ we need to have both $+\langle\varphi\rangle$ and $+\langle\psi\rangle$ and, to introduce $-\langle\varphi \text{ conk } \psi\rangle$ we need to have both $-\langle\varphi\rangle$ and $-\langle\psi\rangle$. Intuitively, this is too much! The conditions for denying φ *conk* ψ should not just be *incompatible* with the affirmation conditions, but should be *minimally* incompatible. Insofar as we need $-\langle\varphi\rangle$ and $-\langle\psi\rangle$ to affirm φ *conk* ψ, needing *both* $-\langle\varphi\rangle$ and $-\langle\psi\rangle$ to deny φ *conk* ψ when *either* by itself is already incompatible with the conditions for affirming φ *conk* ψ violates this minimality constraint. Our task in formulating a criterion of bilateral harmony, then, is to turn these two informal conditions into a pair of formal constraints.

Generalized Bilateral Harmony

The formal constraint corresponding to the first informal condition is obvious. We need to be able to show that, in any case in which opposite stances towards $\varphi \circ \psi$ are introduced and an incoherence is concluded on that basis, the grounds for introducing these opposite stances already suffice to conclude an incoherence without the introduction of opposite stances towards $\varphi \circ \psi$. This amounts to establishing a reduction procedure. For BNK1, the reduction with the first c^* introduction rule goes as follows:

$$
\begin{array}{c}
\dfrac{\dfrac{\mathcal{D}_1 \quad \mathcal{D}_2}{a\langle\varphi\rangle \quad b\langle\psi\rangle}\, c_{\circ\mathrm{I}} \quad \dfrac{\mathcal{D}_3}{a^*\langle\varphi\rangle}}{\dfrac{c\langle\varphi\circ\psi\rangle \qquad c^*\langle\varphi\circ\psi\rangle}{\bot}} \, c^*_{\circ\mathrm{I}_1}
\end{array}
\quad \rightsquigarrow_r \quad
\dfrac{\mathcal{D}_1 \quad \mathcal{D}_3}{\dfrac{a\langle\varphi\rangle \quad a^*\langle\varphi\rangle}{\bot}}\ \text{Inc.}
$$

The reduction with the second c^* introduction rule is analogous. For BNK3, the reduction goes as follows:

$$
\dfrac{\dfrac{\mathcal{D}_1 \quad \mathcal{D}_2}{a\langle\varphi\rangle \quad b\langle\psi\rangle}\, c_{\circ\mathrm{I}} \quad \dfrac{\overline{a\langle\varphi\rangle}^{\,1} \quad \overline{b\langle\psi\rangle}^{\,2}}{\dfrac{\mathcal{D}_3}{\bot}}}{\dfrac{c\langle\varphi\circ\psi\rangle \qquad c^*\langle\varphi\circ\psi\rangle}{\bot}\ \text{Inc.}}\, c^*_{\circ\mathrm{I}}{}^{1,2}
\quad \rightsquigarrow_r \quad
\dfrac{\mathcal{D}_1 \quad \mathcal{D}_2}{\dfrac{a\langle\varphi\rangle \quad b\langle\psi\rangle}{\dfrac{\mathcal{D}_3}{\bot}}}
$$

The reduction for BNK2 is similar. Note that, in the case of *bonk*, there is a combination of c and c^* introduction rules such that no reduction is possible. This formally captures the problem with *bonk* informally articulated above.

It is less obvious as to how the problem with *conk*, informally articulated above, ought to be formally captured. However, I take it that we can arrive at a satisfactory formal constraint by thinking of bilateral harmony by analogy to unilateral harmony. Just as, in the case of unilateral harmony, a reduction shows that the elimination rules are not too strong relative to the introduction rules, and an expansion shows that they're not too weak relative to the introduction rules, in the case of bilateral harmony, a *reduction* shows that it's not *too easy* to conclude opposite stances towards $\varphi \circ \psi$, an *expansion* can show that it's not *too hard* to conclude opposite stances towards $\varphi \circ \psi$. In the case of expansions establishing unilateral harmony, we suppose we have a derivation of $c\langle\varphi \circ \psi\rangle$, we then use the $c_\circ$ elimination rules, making whatever assumptions we must make in order to use them, and derive the grounds required to apply the $c_\circ$ introduction rules and recover $c\langle\varphi\circ\psi\rangle$, having discharged all of our assumptions. Extending this thought analogically, we can think of the application of Incoherence and Reductio, given $c^*\langle\varphi\circ\psi\rangle$, as the application of a kind of elimination rule.[12] This suggests the following expansion procedure. We suppose we have a derivation of $c\langle\varphi\circ\psi\rangle$, and we make whatever assumptions necessary in order to apply the introduction rule for $c^*\langle\varphi\circ\psi\rangle$. Then, using the coordination principles and the introduction rules for $c\langle\varphi\circ\psi\rangle$, we must be able to recover that formula, having discharged all of our assumptions. The fact that we *can* discharge all of our assumptions and reintroduce $c\langle\varphi\circ\psi\rangle$, means that we didn't have to assume *too much* to conclude $c^*\langle\varphi\circ\psi\rangle$. That is, it's not "too hard" to conclude $c^*\langle\varphi\circ\psi\rangle$, relative to how hard it is to introduce $c\langle\varphi\circ\psi\rangle$. Likewise, we suppose we have a derivation of $c^*\langle\varphi\circ\psi\rangle$ and do the same procedure.

The expansions for BNK1 go as follows:

$$\begin{array}{c}\mathcal{D}_1\\c\langle\varphi\circ\psi\rangle\end{array}\rightsquigarrow_e\quad\cfrac{\cfrac{\bot}{a\langle\varphi\rangle}\text{Red.}^1\quad\cfrac{\bot}{b\langle\psi\rangle}\text{Red.}^2}{c\langle\varphi\circ\psi\rangle}c_{\circ I}$$

where the $\bot$'s come from $\cfrac{\begin{array}{c}\mathcal{D}_1\\c\langle\varphi\circ\psi\rangle\end{array}\quad\cfrac{\overline{a^*\langle\varphi\rangle}^{\,1}}{c^*\langle\varphi\circ\psi\rangle}c^*_{\circ I_1}}{\bot}\text{Inc.}$ and $\cfrac{\begin{array}{c}\mathcal{D}_1\\c\langle\varphi\circ\psi\rangle\end{array}\quad\cfrac{\overline{b^*\langle\psi\rangle}^{\,2}}{c^*\langle\varphi\circ\psi\rangle}c^*_{\circ I_2}}{\bot}\text{Inc.}$

$$\begin{array}{c}\mathcal{D}_1\\c^*\langle\varphi\circ\psi\rangle\end{array}\rightsquigarrow_e\quad\cfrac{\bot}{c^*\langle\varphi\circ\psi\rangle}\text{Red.}^3$$

with the intermediate steps using $\cfrac{\overline{a\langle\varphi\rangle}^{\,1}\quad\overline{b\langle\psi\rangle}^{\,2}}{c\langle\varphi\circ\psi\rangle}c_{\circ I}$, $\cfrac{\overline{a^*\langle\varphi\rangle}^{\,1}}{c^*\langle\varphi\circ\psi\rangle}c^*_{\circ I_1}$, $\cfrac{\overline{b^*\langle\psi\rangle}^{\,2}}{c^*\langle\varphi\circ\psi\rangle}c^*_{\circ I_2}$, and Inc./Red. steps as shown.

The expansions for BNK3 go as follows:

---

[12] Of course, coordination principles can also be likened to introduction rules (cf. Kürbis (2021)). The reason for thinking of likening coordination principles to elimination rules here is simply that, following Gentzen's principle of prioritizing introduction rules in the context of proof-theoretic semantics approach, the criterion of bilateral harmony is formulated for positive and negative introduction rules.

$$
\begin{array}{c}
\mathcal{D}_1 \\
\boldsymbol{c}\langle\varphi \circ \psi\rangle
\end{array} \rightsquigarrow_e
\quad
\dfrac{\overline{\boldsymbol{a}\langle\varphi\rangle}^{\,1} \quad \overline{\boldsymbol{a}^*\langle\varphi\rangle}^{\,2}}{\quad} \text{Inc.}
\qquad
\dfrac{\overline{\boldsymbol{b}\langle\psi\rangle}^{\,3} \quad \overline{\boldsymbol{b}^*\langle\psi\rangle}^{\,4}}{\quad} \text{Inc.}
$$

(Expansion diagrams for $\boldsymbol{c}\langle\varphi \circ \psi\rangle$ and $\boldsymbol{c}^*\langle\varphi \circ \psi\rangle$ using reductions Red. and Inc., with labels $\boldsymbol{c}^*_{\circ\mathrm{I}}{}^{1,0}$, $\boldsymbol{c}^*_{\circ\mathrm{I}}{}^{3,0}$, $\boldsymbol{c}_{\circ\mathrm{I}}$, and $\boldsymbol{c}^*_{\circ\mathrm{I}}{}^{1,2}$.)

$$
\begin{array}{c}
\mathcal{D}_1 \\
\boldsymbol{c}^*\langle\varphi \circ \psi\rangle
\end{array} \rightsquigarrow_e
\quad
\dfrac{\overline{\boldsymbol{a}\langle\varphi\rangle}^{\,1} \quad \overline{\boldsymbol{b}\langle\psi\rangle}^{\,2}}{\boldsymbol{c}\langle\varphi\circ\psi\rangle}\,\boldsymbol{c}_{\circ\mathrm{I}}
\qquad
\begin{array}{c}\mathcal{D}_1 \\ \boldsymbol{c}^*\langle\varphi\circ\psi\rangle\end{array}
$$

The expansions for BNK2 are similar. No such expansions are possible in the case of *conk*.

I submit that this criterion of bilateral harmony provides both necessary and sufficient conditions for a set of positive and negative rules being harmonious. On the one hand, it is satisfied by BNK1, BNK2, and BNK3, three systems whose rules are intuitively harmonious. On the other hand, it rules out *bonk*, *conk*, and every other bilaterally dissonant connective that has been proposed in the literature. Moreover, it really does get to the heart of the matter as to what is wrong with these connectives.

6 Conclusion

In this paper, I've done three main things: (1) I've presented, in generalized fashion, three bilateral natural deduction systems for classical logic, (2) I've provided a new criterion for bilateral harmony that I argued is superior to existing criteria that have been proposed in the literature, and (3) I've shown that all three systems meet it. I'll briefly conclude with three directions for further work.

First, I have argued that it is a virtue of my proposed criterion of bilateral harmony that all three intuitively harmonious systems meet it. A consequence of this, however, is that a bilateral natural deduction system's being both unilaterally and bilaterally harmonious does not suffice to establish that its rules are uniquely definitive of the meanings of the classical connectives, since there are multiple such sets of rules. So, one has a choice: either supplement harmony with some further proof-theoretic constraint to distinguish one set of rules as uniquely definitive or give up on the idea that there is some one such set of rules. If one opts for the first option, then one such constraint, giving some grounds to prefer BNK2 or BNK3 over BNK1, may be that proposed by del Valle-Inclan and Schlöder. However, some further grounds will be needed to decide between BNK2 and BNK3. If one goes in for the second option, then one needs to say, what, exactly, a proof-theoretic specification of the meanings of the classical connectives comes to if not the specification of a set of rules that define the meanings of the connectives. Either way, there is work to be done.

Second, though my formulation of bilateral harmony is novel in the context of bilateral natural deduction systems, as is well-known, there is a very close correspondence between bilateral systems and multiple conclusion sequent calculi. While there is not the space here to develop this claim, my two constraints correspond very closely to two constraints often placed on multiple conclusion sequent calculi in the context of proof-theoretic semantics (Hacking (1979), Kremer (1988)): the eliminability of Cut and the eliminability of non-atomic instances of the Identity axiom. I've developed this same basic conception of bilateral harmony, under this different guise, elsewhere (Simonelli (2024a)). A worked-out account of the relation between these two approaches would be illuminating.

Finally, I have restricted my attention here to *classical* logic, as this has been the main focus of developments in bilateralism following Smiley and Rumfitt. However, there has been developments of bilateral natural deduction for non-classical logics in recent years.[13] This general schematized approach to bilateralism as well as the more specific approach to bilateral harmony may be fruitfully applied to such developments.[14] So, though my focus here has been classical logic, the menu of harmonious bilateral systems given in this paper, the generalized approach through which they've been stated, and the method for establishing bilateral harmony will likely be of use to those looking to apply bilateralism beyond classical logic.

---

[13] See note 2 above for some such developments. See also Francez (2014b), Francez (2023), and Simonelli (2024b).

[14] In this regard, it is important that the criterion of bilateral harmony proposed here is a more permissive one than those that have been proposed in the literature, since, while the BNK2 and BNK3 rules are suitable for classical logic, they will not be suitable for many non-classical logics.

References

Ayhan, S. (2021). Uniqueness of logical connectives in a bilateralist setting. In M. Blicha & I. Sedlár (Eds.), *The Logica Yearbook 2020* (pp. 1–16). London: College Publications.

Brandom, R. (1994). *Making It Explicit: Reasoning, Representing, and Discursive Commitment*. Cambridge, Mass.: Harvard University Press.

del Valle-Inclan, P. (2023). Harmony and normalization in bilateral logic. *Bulletin of the Section of Logic, 52*, 377–409.

del Valle-Inclan, P., & Schlöder, J. (2023). Coordination and harmony in bilateral logic. *Mind, 132*, 192–207.

Drobyshevich, S. (2019). Tarskian consequence relations bilaterally: Some familiar notions. *Synthese, 198*(S22), 5213–5240.

Ferreira, F. (2008). The co-ordination principles: A problem for bilateralism. *Mind, 117*(468), 1051–1057.

Francez, N. (2014a). Bilateralism in proof-theoretic semantics. *Journal of Philosophical Logic, 43*, 239–259.

Francez, N. (2014b). Bilateral relevant logic. *Review of Symbolic Logic, 7*(2), 250–272.

Francez, N. (2015). *Proof-Theoretic Semantics*. London: College Publications.

Francez, N. (2023). Bilateral connexive logic. *Logics, 1*(3), 157–162. Retrieved from https://www.mdpi.com/2813-0405/1/3/8

Gabbay, M. (2017). Bilateralism does not provide a proof theoretic treatment of classical logic (for technical reasons). *Journal of Applied Logic2, 25*, S108–S122.

Gentzen, G. (1935). Untersuchungen über das logische Schließen. I. *Mathematische Zeitschrift, 35*, 176–210.

Hacking, I. (1979). What is logic? *Journal of Philosophy, 76*(6), 285–319.

Hjortland, O. T. (2014). Speech acts, categoricity, and the meanings of logical connectives. *Notre Dame Journal of Formal Logic, 55*(4), 445–467.

Incurvati, L., & Schlöder, J. J. (2023). *Reasoning with Attitude*. New York: Oxford University Press USA.

Kremer, M. (1988). Logic and meaning: The philosophical significance of the sequent calculus. *Mind, 97*(385), 50–72.

Kürbis, N. (2021). Normalization for bilateral classical logic with some philosophical remarks. *Journal of Applied Logics, 8*, 531–556.

Kürbis, N. (2022). Bilateral inversion principles. *Electronic Proceedings in Theoretical Computer Science, 358*, 202–215.

Murzi, J. (2020). Classical harmony and separability. *Erkenntnis, 85*(2), 391–415.

Pfenning, F., & Davies, R. (2001). A judgmental reconstruction of modal logic. *Mathematical Structures in Computer Science, 11*(4), 511-540.

Prawitz, D. (1965). *Natural deduction: A proof-theoretical study*. Mineola, NY: Dover Publications.

Prior, A. (1967). The runabout inference ticket. In *Analysis* (pp. 38–9).

Rumfitt, I. (2000). "Yes" and "no". *Mind, 109*, 781–823.

Simonelli, R. (2024a). A general schema for bilateral proof rules. *Journal of Philosophical Logic, 3*, 623–656.

Simonelli, R. (2024b). *"Yes," "No," "Neither," and "Both": Bilateral Systems for the FDE Family*. Retrieved from `https://www.ryansimonelli.com/papers.html`

Smiley, T. (1996). Rejection. *Analysis, 56*, 1–9.

Smullyan, R. M. (1968). *First-Order Logic*. New York: Springer Verlag.

Wansing, H. (2016). Falsification, natural deduction and bi-intuitionistic logic. *Journal of Logic and Computation, 26*, 425–450.

Wansing, H. (2017). A more general general proof theory. *Journal of Applied Logic, 25*, 23–46.

Wansing, H., & Ayhan, S. (2023). Logical multilateralism. *Journal of Philosophical Logic, 52*(6), 1603–1636.

Ryan Simonelli
Wuhan University, School of Philosophy
China
E-mail: `ryanasimonelli@gmail.com`

Logic as Liberation, or, Logic, Feminism, and Being a Feminist in Logic

SARA L. UCKELMAN[1]

Abstract: There has been a long history of tension between feminists and feminist philosophy, on the one hand, and logic, on the other hand. This tension expresses itself in many ways, including claims that logic is a tool of the patriarchy, that logic/rationality/analytical tools in philosophy need to be rejected if women are to fully participate, that woman = body and man = mind, that to do feminist philosophy one must do it as a situated, embodied person, not as an impersonal, disembodied mind, that logic is "a masculine subject". However the tension is expressed, it is women in logic and women logicians who are caught in between. The goal of my paper is to explore a conception of logic that not only is not inconsistent with being a feminist, but is actively welcoming of women as logicians.

Keywords: feminist logic, feminist philosophy, Andrea Nye, Val Plumwood

1 Introduction

There has been a long history of tension between feminists and feminist philosophy, on the one hand, and logic, on the other hand. This tension expresses itself in many ways, including claims that logic is a tool of the patriarchy, that logic/rationality/analytical tools in philosophy need to be rejected if women are to fully participate in the field, that logic is "a masculine subject" (Nye, 1990, p. 2), that "woman = body" and "man = mind," that women have "brought the body into philosophy," that to do feminist

---

[1] The author is grateful to the welcoming and supportive audience of Logica 2023, where this material was first put into words. She would also like to thank the enthusiastic support of the Durham Undergraduate Philosophy Society, who were treated to a more refined version of this material in spring 2024.

philosophy one must do it as a situated, embodied person, not as an impersonal, disembodied mind.[2] Logic is furthermore too often used as a tool of oppression, a means to challenge, disenfranchise, and demean women (and others!) as too emotional, too irrational.

The anti-feminist conception of logic outlined above is espoused by people like Andrea Nye, especially in her book *Words of Power* (Nye, 1990). This conception is rooted in binaries: logical vs. non-logical, male vs. female, man vs. woman, logician vs. feminist. We will show that to follow Nye, and agree that logic is the purview of men, and not of women, is to buy into a problematic story.

Val Plumwood has challenged this approach by highlighting problems that arise from viewing the world through these dualisms, which are "a particular way of dividing the world which results from a certain kind of denied dependency on a subordinated other" (Plumwood, 1993, p. 443). Because of the dominating nature of these dualisms, it is the incorporation of dualisms into logic that is problematic for using logic to achieve feminist aims, Plumwood argues, not logic itself. The focus of her concern is the dualistic nature of classical logic, which divides the world into "true" and "not true". As an alternative, she argues for the incorporation of relevant negation (Plumwood, 1993, p. 458), which does not have the same hierarchical, homogenizing effect that classical negation has.

Plumwood's article provides us with a model for how we can begin to understand logic in a non-anti-feminist way. But Plumwood herself does not challenge the broader societal binaries that are still attached to logic in problematic ways that impact on the recruitment and retention of women in logic. This makes challenging the gender binary of man vs. woman, with the dependent binaries of logical vs. emotional and mind vs. body, an integral step of making logic an inclusive place for not only women, but also everyone who falls outside of the gender binary. Instead of merely rejecting the problematic identification of man = logic and woman = emotion, we should be rejecting the binary between man and woman altogether; without this binary, all binaries dependent on it fall apart.

However the tension is expressed, it is women in logic and women logicians who are caught in between. The goal of this paper is to explore a conception of logic which is not only consistent with being a feminist, but is actively welcoming of women as logicians. The aim will be to justify a

---

[2]I'm not claiming that *all* feminists adopt these views; but each of these sentiments are ones that I have had expressed to my face from people who claim to be feminists or to support women in logic and/or philosophy.

Logic as Liberation

position where logic is no longer a tool of oppression and domination but is instead a tool of liberation.

The paper is both theoretical and autobiographical. This is because we cannot judge the impact on lived experiences of how society approaches and conceives of topics such as gender and rationality without examples of these experiences. While I do not wish to claim that my experiences are universally generalizable, I share them because I know that aspects of them will resonate with readers in different ways, whether because they have had similar experiences or whether because they have witnessed people having similar experiences.

In the next section, I address the important question of *why does this matter?* Where do these concerns come from? Who cares whether logic is a tool of oppression or a tool of liberation? In sections 3 and 4, I look at two historical accounts of the relationship between logic and women, both of which illustrate the ways in which logic has, historically, been set up as a tool to repress or exclude women—a phenomenon which is the focus of section 5. In section 6, I offer a positive account, of how logic can be used as a tool of liberation rather than exclusion.

2 Why does this matter?

In late January 2023, my labor union (the University and College Union) announced 18 strike days across February and March. The impact of this industrial action was that my introductory logic class went from having eight lectures left to having *three* (on top of having already lost one lecture the previous term due to strikes). Of these three lectures, one was the day after the announcement, with a topic that couldn't be changed at such short notice.

It was heartbreaking.

There was no way that I could teach all the topics I normally teach with the loss of five lectures. I had to answer a crucial question: If I only had two weeks left, what is it that I wanted my students to learn? Ordinarily, these eight lectures would have involved finishing up the meta-theory of propositional logic, segueing into predicate logic semantics and proof theory (and a brief discussion of meta-theory), and then we'd wrap up the year by shifting gears entirely to look at Buddhist logic.

I decided that if I had to lose a large chunk of material, it was predicate logic that had to go, in favor of keeping as much of my usual two lectures on Buddhist logic as I could. But this raised another question: How would

I explain this to my students? Not just that the Buddhist logic material is important, but that it was *more important* that they learn about it than that they learn predicate logic. And how could I explain to them, too, why this was even an issue—why, despite the fact that there is literally nothing that I enjoy doing more than teach intro logic to enthusiastic students, I would be participating in industrial action, even with the disastrous effect it would have on my favorite activity. If I wanted to give one lecture on Buddhist logic, then this meant I had *one* lecture to get them to understand why any of this mattered.

So when my lecture came around the next week, instead of talking about the language of predicate logic, I told my students we were going to talk about why we were doing this: Why are we even in this class? We covered possible answers all the way from "so that you can achieve the subject-specific learning outcomes listed in the module description as published in the faculty handbook" to "I hope you learn what logic is and what logicians do" to "how to follow rules/directions and reason from a definition" and then we had a collective discussion on the bigger questions:

- Who is logic for? Who gets to be or count as a logician? Who is excluded?

- What are ways in which "logic" or "reason" or "rationality" (especially claims of "being reasonable" or "being rational") weaponized in modern Western society? Who does this weapon tend to be used against?

- If we are currently living in a society that is under the "rule of reason", what would an alternative to this rule look like? Could reason/rationality/logic still play a role?

I had never explicitly discussed these questions with students before, and underestimated the impact they would have. Most of the students were in the class because they were interested in logic, rather than because they are interested in wider social issues, but *all* of them, that day, recognized that our study and use of logic doesn't exist in a vacuum, and that these questions, far from being irrelevant wokery, should be central to both the study and the practice of logic.

One of the main purposes of this paper is to illustrate *why*.

3 Women and logic

3.1

I went into my first logic class convinced I was going to fail.

I was still in high school, and had signed up to do a class at the local community college. Why logic? Because I'd lived all my life knowing, fundamentally, to my core, that I was not a logical person—and that my father was. I wanted to know more about how he thought, how he worked, and this was the perfect opportunity to do so, because the grades I got wouldn't count towards anything. It wouldn't matter when I failed.

And then, I didn't. About a third of the way through the semester I realized I was the only one who understood any of what was going on; and by the end of the semester, I had tutored all but one of the other students. Not only was I *not* fundamentally illogical, I was actually rather good at it!

By the time I graduated university, I had taken all of the logic classes offered in both the philosophy and the mathematics departments (including one that I took twice!). Logic was everything that English literature (my first major) and philosophy (my second major) were not: There were rules. Things were right or wrong. Either you had a proof or you didn't. Questions could actually be answered; and the task of figuring out the right new questions to ask was enormously challenging and satisfying. Beyond this, logic also gave me the tools I needed to structure my world into something I understood. It gave me tools for navigating social situations, to make decisions about my future in the face of uncertainty, and to be a better parent. It has also given me the opportunity to share the sheer joy of it with others.

3.2

In 1913, American author, historian, and Unitarian minister Edward E. Hale published an article in *The North American Review* entitled "Women and Logic."[3] It opens:

> That women are not logical is one of the recognized conventions of social life (Hale, 1913, p. 206).

In his article, Hale is interested in two questions: (1) Where this convention comes from/what underpins it, and (2) What, exactly, is meant by the convention.

---

[3] For quite some time—but thankfully no longer—this article was Google's top hit when searching for "women in logic."

To the first question, Hale says he only knows of one explicit discussion of the convention, in Otto Weininger's *Sex and Character*, which he attempts to summarize. In brief: A necessary part of the logical faculty is memory—in order to be able to carry out logical operations, one must remember what you have started with, and what steps one has done along the way, so that the A that you start with is the A that you end up with. Memory is also required in order for any sort of generalization across time to be made: One has to be able to remember the instances yesterday in order to recognize that they are the same as instances today, and thus that these instances might be instances of some more general law. "Only so can we understand the fundamental proposition of Logic, $A = A$", Hale (1913, p. 206) says. But, according to Weininger, "the absolute woman has no memory" (Hale, 1913, p. 206). So she lacks one of the necessary components of the logic faculty.

Hale notes that Weininger's position can be objected to by either rejecting the claim that memory is necessary for logic, or the claim that women do not partake in memory; he prefers to avoid both of these questions and instead "examine the general proposition directly; make a frontal attack, as one might say" (Hale, 1913, p. 207). This leads him to the second question, namely, what does it mean to say that women are not logical? The answer to this, says, depends on what we mean by "logic" (Hale, 1913, p. 212). He identifies four possibilities:

1. the logic of the schools

2. the logic of argument

3. the logic of consistency

4. the consistency between theory and practice

These are not the only possibilities, but he notes that he sets aside from consideration many interpretations of "logical" on which not only are women not logical, but men are not either.

Even under the four possibilities that he focuses on, Hale admits that women do not have the monopoly on being illogical. By the "logic of the schools", Hale means the formal study of logic, via the reading of textbooks and the discussion of arguments in semi-formal representation In this "scholastic sense," "most men are not logical... they have no idea even of what logic is" (Hale, 1913, p. 207), because they have never been exposed to the formal mechanisms of logic. Nevertheless, they are still able to spot fallacies and reason according to logical principles; but this—Hale admits—is

Logic as Liberation

something that women are able to do to: "Women are just like men in this respect" (Hale, 1913, p. 208).

The next alternative Hale considers is the logic of argument, that is, the practical application of correct reasoning, whether based on the logic of the school textbooks or on commonsense principles; to say then that men are logical (and women are not?) is to say that they are "particularly argumentative" (Hale, 1913, p. 210). But he notes this account of "logical" occupies only a small portion of our every day use of the term. Instead, quite often, "logical" is used in a non-argumentative way, to mean that there is "a sense of consistency of coherence, [...] a feeling of what is necessary, of how a matter ought to turn out, of what is proper" (Hale, 1913, p. 210). Logic, then, is "a matter of demonstration and of proof" (Hale, 1913, p. 211), resulting in conviction. If we take "logical" in either of these, every day, senses, then women *are* illogical, in the sense that "when they try to prove anything they come out at an illogical result or they get at their result by illogical methods" (Hale, 1913, p. 211) because they "argue not by making inferences or deductions, but determine their result by intuition or by some other method known to themselves" (Hale, 1913, p. 211).

This brings Hale to the point of being forced to grudgingly concede that maybe women do have some little bit of logic, perhaps, if we construe logic in the right way, if it means "consistency [...] with whatever plan, good or bad, happens to be under discussion" (Hale, 1913, pp. 216–217) or "a sort of consistency or coherency, a full development or a natural outcome or something of the sort" (Hale, 1913, p. 211)—if we do this, "then we shall often find that these intuitions of women are often logical enough" (Hale, 1913, p. 211) (even if this has "nothing to do with logic considered as argument" (Hale, 1913, p. 211)).

A munificent conclusion! Maybe 16-year-old me would've been reassured, in advance of taking my first logic class, that when I questioned whether I could be logical, I could reassure myself that as a woman, I did have some "logic"!

But the real question is not "whether women can be logical" but *how this is even a question at all*??

Sara L. Uckelman

4 Woman *or* logic?

4.1

There is something shocking, to the 21st-century reader, in the opening lines of Hales' paper, a visceral gut-punch of *how can this even be a question?* Surely, by now, we might hope, more than a century after Hale was writing, *things have gotten better*, that we can take it for granted that being a woman and being logical are not incompatible.

We might hope, and our hopes will be dashed.

I left my initial PhD programme, in philosophy, because I wanted to do logic but was told that logic was "not philosophical enough." Since then, I have had an uneasy relationship with being labeled a philosopher. In 2016, I had the pleasure of going to Australia for a string of back to back conferences—the Australasian Association for Logic; the Australasian Association for Philosophy; and the International Association for Women Philosophers. This series of conferences really drove home to me how this reluctance to include logic within the purview of philosophy goes both ways.

I went from feeling included and welcome and within my element, at the AAL, to feeling ostracized and unwanted, at the IAPh. The clincher event was a roundtable on Women and Philosophy, where the idea that one could be a woman, in philosophy, who was interested in logic and used logical tools was an anathema. I came away from this roundtable with a profound feeling that if this is what 'Women and Philosophy' is, then the only logical conclusion is that either I am not a woman (or the right kind of woman) or what I do is not philosophy (or not the right kind of philosophy).

4.2

Fastforward almost 80 years from Hale: Far from us having reached the enlightened position where no one even questions whether women can be logical, we find ourselves in a position where people—with explicitly feminist leanings!—reject the suitability of logic for women entirely:

> Logic... is not a feminine subject... logic is, after all, a masculine subject (Nye, 1990, p. 2))

Additionally, according to Nye (1990, p. 5), "logic articulates oppressive thought-structures that channel human behavior into restrictive gender roles." This position goes a step beyond creating a concept of "women's logic" as Hale does and simply denies that women can have any logic at all. Nye excludes *a priori* any possibility of being both a woman (or a feminist!) and a logician. This is because "logic is the creation of defensive male subjects who have lost touch with their lived experience" (Nye, 1990, p. 4) and "an invention of men, that is something men do and say" (Nye, 1990, p. 5). The exclusion is affirmed again at the end of the book, when she consider the case where "a feminist reader is to remain reader *and not turn logician at the last moment*" (Nye, 1990, p. 175, emphasis added) (never pausing to wonder if one could be both reader and logician at the same time).

As a logician, it is hard to read a critique such as Nye's because I want to *also* be able to be a feminist. If Nye is right, no woman can be both a feminist and a logician together, because the two are antithetical to each other. If she is right, then if I wish to be both feminist and logician, then—because I am a logician and can do modus tollens—I must conclude that I cannot be a woman.

Or, we can find ways to resist Nye's dogmatic conclusions.

Nye's arguments are rooted in her own experiences, as the introduction to her book makes clear. She is at pains to be explicit about how her reading of the history of logic is influenced by her position as "a woman reading logic" and "a philosopher who, like many other women philosophers, has often felt uneasy claiming that title" (Nye, 1990, p. 5), and how perhaps it is "*only* a woman...a woman uncomfortable in the world of men...a woman too intent on emotional commitments to be capable of purely abstract thought" (Nye, 1990, p. 5) that can make the arguments she makes. More importantly, she emphasizes how "there is never only one read*ing*" [emphasis added] or truth about the nature of logic (Nye, 1990, p. 5), because there is never only one read*er*. Nevertheless, she takes her experiences as generalizable, and bases sweeping conclusions about the gendered nature of logic on them. If we start from a different set of experiences, we could very easily be lead to different readings, different conclusions.

My own experience with logic has been almost entirely opposite to Nye's. Instead of logic being a place where I am forced into a particular gender role, it is the place where I first felt freed from restrictive gender roles, where I could escape the woman = emotion/woman = body/woman = irrationality/men = rationality equations that had pervaded my life, implicitly and explicitly. Logic was where I first felt at home, a place within philosophy where my gender didn't matter. Logicians have never judged my desire to be a logician on the basis of my gender.

How are we to reconcile two such different experiences? Without taking anything away from Nye's conclusion that logic is not the subject for her,

Sara L. Uckelman

I would like to explore reasons why two people of the same gender could experience the same subject in two such very different ways.

One thing that has always drawn me to logic is that it is a rule-following discipline. The rigidity of the rules is part of its attraction: You either have a proof, or you don't. Something is either true, or it is false. (Or both, or neither, but whichever combo it is, it is clearly and definitively). Sometimes your answer is wrong; sometimes it is right. And when something is wrong, or false, or not a proof, there is something that you can say which explains why. This is part of what also makes logic such a joy to teach, because there is always a reason why students have gotten something wrong, and if I can help them see that reason, then there's nothing to prevent them from fixing it and never making that mistake again.

It's only been in recent years that I've realised how connected my joy in logic is to my joy in the slavish adherence to arbitrary rules, and how much *that* is an indicator of autism. The deep-seated need to be a rule-follower (combined with a rampant desire to question everything, including the rules I want to follow!) along with an inability to deviate from the rules or to understand how other people can just blow those rules off, is one of the most central aspects of my character, and one of the strongest signs that I have that I am autistic. While I do not believe any research has been done specifically into the prevalence of diagnoses, or suspected diagnoses, of ASD within the community of logicians, anecdotal evidence indicates that the prevalence within the logical community is higher than in the wider community. This is of no surprise, that people who delight in rules and structure and clarity would be attracted to a field that is designed around these very factors!

What is perhaps more surprising is the way that these factors intersect with questions of gender. While research into the connection between gender identity or gender expression and autism is still in its early stages, recent studies have demonstrated a strong overlap between people who have atypical gender identities and people who are on the autistic spectrum (Cooper, Smith, & Russell, 2018; Lai, Lombardo, Auyeung, Chakrabarti, & Baron-Cohen, 2015; van Schalkwyka, Klingensmith, & Volkmar, 2015), with there being a higher correlation between gender variance in autistic people than in non-autistic people (van Schalkwyka et al., 2015, p. 81). This is especially pronounced amongst autistic people who are assigned female at birth, who experience this gender variance at a higher rate than autistic people who are assigned male at birth (Cooper et al., 2018, p. 3995).

The relationship between gender identity and autistic identity is complex and operates at many levels. The diagnostic criteria for autism is centered around how it is manifest in male-presenting subjects, which can result in "a diagnosis of autism, with traits perceived to be male, [having] implications for how autistic natal females feel about their biological sex and gender" (Cooper et al., 2018, p. 3995). The gendered nature of the criteria for diagnosis also results in women being "less likely to be diagnosed with ASD than boys despite demonstrating similar levels of autistic symptoms" (Duvekot et al., 2017, p. 646). It is only in the last decade and a half that researchers have specifically addressed the issue of gender differences in autism diagnoses, and this only happened "a result of women themselves questioning their often late diagnosis" (Gould, 2017, p. 703).

One final point on the nature of the varied experiences of women in logic: What does it mean to call a subject or a discipline *masculine*? Nye takes it to be the fact that logic is, predominantly, a field in which men are active, and from this she draws conclusions that seek to *exclude* women. But this is to take a descriptive position and draw a normative conclusion from it, when one can just as easily draw a descriptive conclusion: If I am a feminist, and I am doing logic, then what I am doing is "feminist logic." Similarly, if logic is to be considered a part of philosophy, then if I am a woman, and I am doing logic, then what I am doing is "what women in philosophy do." All of this to say: It is not clear that there is any value to be gained from demarcating logic as a "masculine" subject, one that is antithetical to either feminism or being a woman. Attempts to gender the field of logic range from being misguided to being simply irrelevant.

So if you ask me "are you a man or a woman?" I will answer "I am a logician".

5 The weaponization of logic

5.1

I asked my students to tell me, if they felt comfortable sharing this in a full lecture hall, how many of them had ever been told "don't be so emotional" or "be more rational," as a means of shutting down or removing them from the discourse. I was not surprised how many people raised their hands; women, at least, have been told this all our lives.

I was surprised, and saddened, by how many men did.

This is logic being used as a weapon, as a tool of silencing. It is one person telling another: "I will only interact with you on my terms, not yours."

It is a way of denying other people a voice, a way of saying that they are not full participants in human discourse.

How can we keep logic from being weaponized?

5.2

I, as a logician, clearly think that logic is something valuable, the study and application of which is worth pursuing myself and worth teaching others to want to pursue. How do I know that I am not mistaken in thinking this? How do I ensure that I am not participating in oppression by promoting, using, and teaching logic? To answer these questions, we must take Nye's claim that logic is a tool of oppression or a tool of the patriarchy seriously. If it is, two questions arise: (1) How is it used as an oppressive tool? and (2) How can we mitigate this use?

One way in which logic can be used as an instrument of oppression is by using it as a demarcation of who gets to count: Whose voices are we going to listen to, in debate and conversation both public and private? When someone is told that they are not "being rational" or "being logical" in their participation in a conversation, this is not an invitation to change their behavior, but it is a means of ending the conversation or forcing them out of the discussion. Examples of this silencing can be found in contemporary pundits such as Rush Limbaugh and Jordan Peterson, but the weaponization of logic to exclude certain voices from the discussion isn't merely a modern conceit, and it isn't (and hasn't been) used only against women. As Plumwood points out, decisions about who gets to be within the sphere of reason and who is excluded from it go at least as far back as Plato and Aristotle, and

> For Kant, it is not only women who are excluded from reason by their possession of a gallantly presented but clearly inferiorised 'beautiful understanding', but also workers, and blacks (Plumwood, 1993, p. 436).

The way in which logic, logicality, reason, and rationality can be used as an instrument of oppression goes beyond merely shutting people outside of conversation; it is also possible for it to be uses as a means of *denying* personhood to people in the first place. If we take seriously Aristotle's definition of humans as "rational animal," then by charging people, individually or collectively, as "irrational," it is possible to deny them their essential humanity. This goes beyond just "who are we going to listen to" or "who gets to speak" to "who gets to be counted even when others speak". Who is going to get to count as "rational," and hence of value?

In 1824, Henry T. Colebrooke presented an account of the Nyāya system of logic to the Royal Asiatic Society. This was the first time that European philosophers were exposed to non-European systems of logic (Ganeri, 2001, pp. 4–5). Such exposure was more than just the opening of an interesting field of inquiry to rarified academics: Rather, as Ganeri (2001) convincingly argues, the idea that there were "subject matter ideas and theories closely akin to those of the Greek founders of Western philosophy" [p. 5] within the corpus of Indian philosophical texts threatened one of the basic axioms of Western colonization of the East:

> The assumption that the West, and the West alone, had developed a science of reason was a fundamental axiom in the justification of the colonial enterprise as a civilizational process (Ganeri, 2001, p. 4).

If, as Aristotle says, man is "rational animal," and if Western civilisation (aka, civilisation) is built on its inheritance of Greek philosophical thought, including Aristotle's logical corpus, then any culture that was not built upon this same foundation would not have the same access to civilisation. The colonist's argument then goes: If we think that being civilized is a good thing, then we should be seeking to bring civilisation to other cultures. The discovery of a structured system of reasoning that had independent roots from Aristotle threatened the early nineteenth-century British colonial programme by undermining the basic principle that the West had a monopoly on principles of reasoning and rationality. It threatened "the European self-understanding of its intellectual superiority over its colonies" (Ganeri, 2001, pp. 5–6), the very premise upon which colonizers were able to justify their colonial practices.

It is hard to escape Nye's conclusion that logic *can* be used as a tool of oppression. Given that, what can we do to reject its use as this type of tool— and can we even do so? Nye clearly thinks that there is no rehabilitating logic; its use as a tool of domination is all pervasive and inescapable. Plumwood, on the other hand, rejects Nye's broad brush characterization of logic as *inherently* problematic, and asks:

> Why does it make a case for abandoning logic, as opposed to critically reconstructing it and making much more limited claims for it? (Plumwood, 1993, p. 438).

Sara L. Uckelman

Plumwood argues that Nye has adopted an overgeneralized and overabstract account of logic, created from the very logical structures that Nye wishes to reject. Instead, if we take a more nuanced approach to what logic is—and what it can be—we can provide space for a more positive account of logic.

Plumwood locates the problem with logic not in logic itself but in the way in which the core of contemporary logic, and the dominant tradition throughout Western history, namely classical logic, buys into a system of dualisms; it is these dualisms that make logic apt for oppression. The specific dualism that Plumwood identifies as the core problem is the dualism of negation: Classical negation is a tool of oppression, according to Plumwood, because of its dualistic nature forces us to carve up the world into X's and non-X's — true and not-true, man and not-man, human and not-human, rational and not-rational. This way of thinking supports "the structure of a general way of thinking about the other which expresses the perspective of a dominator or master identity, and thus might be called a logic of domination" (Plumwood, 1993, p. 442). It is the homogenizing effect of classical negation that Plumwood identifies as problematic: When carving the world up into X's and not-X's, it is one thing to be an X, but it is many things to be a non-X. It is one thing to be a man; but to be a non-man is to be a woman, or an enby, or any of many other things. The X/non-X binary erases all the differences in the non-X category, defining its members in terms of what they are not rather than in what they are. To paraphrase Tolstoy, all Xs are alike; but every non-X is not X in its own way. Yet classical logic erases all of these distinctions under a single label, "non-X".

But such a dualistic approach to negation, and othering, is not an intrinsic part of logic; it is a choice that logicians make:

> The 'naturalness' of classical logic is the 'naturalness' of domination, of concepts of otherness framed in terms of the perspective of the master (Plumwood, 1993, p. 454).

That is, it is no more 'natural' than domination itself is. There are other accounts of negation which do not carve up the world in this dualistic way, and these logics are less able to be used as tools of oppression and domination. Since we can have logic without classical negation, Plumwood argues that Nye errs in rejecting logic completely. Instead we should reject the binary enforced by classical negation, but keep logic while adopting another type of negation (she argues for something relevant (Plumwood, 1993, p. 458); for a full discussion of how Plumwood uses relevant logic for feminist purposes, see (Eckert & Donahue, 2020)). When logic takes into account not the dualistic division of the world into Xs and non-Xs, but rather the differences that make up the non-Xs, then we are in a position to develop, as Plumwood (1993, p. 458) says, a "liberatory logic"—the logic that will set us free, rather than bind us.

6 Logic as a tool of liberation

What would logic look like, as a tool for liberation? Plumwood (1993, p. 439) says that we must "really insist that all uses of language be grounded in personal experience, the testimony of the witness, and 'the normality of human interchange that logic refuses'." Rather than allowing binary dualisms to erase important differences, we should embrace these differences and make them explicit. If logic is properly grounded in careful attention to personal experience, testimony, human interchange, etc., then logic no longer has to be a tool of domination:

> If there is not one Logic, but in fact many different logics, if logics can be constructed which can tolerate even contradiction itself, logic itself can have no silencing role and no unitary authority over language (Plumwood, 1993, p. 440).

Instead of being used as a means of silencing, properly deployed it can be a means of giving a voice to the voiceless. In this concluding section, I would like to give an illustration of one way that this can be done.

As I foreshadowed in the introduction, this paper is partly theoretical and partly autobiographical. I bring in the autobiography precisely because it is an example of that personal experience and testimony that provides us with an exemplar of the liberatory role that logic can play. Let me return to two points discussed above, namely, the experiences that have made clear that my enjoyment in logic is an aberration, that it makes me "the wrong kind of woman," on the one hand, and the connection I drew between what draws me to logic and aspects of myself as an autistic person, on the other hand.

As an autistic person, gender, for the most part, is simply incidental to me, a non-thing. I sometimes joke that I am "cis by default," because being anything else would simply be too much work. I can see why the idea of being agender would be attractive; but it also seems like a big hassle, when things have gone along fine enough so far. So I am cis, I am a woman, and I am a logician. Maybe I am "the wrong kind of woman", but if so, it is precisely because my view of my gender is mediated through my autism, which itself is intimately linked to my love of logic. The research discussed above, on the relationship between gender identity and neurodivergence, shows that, especially for autistic women, one cannot disentangle one's experience of autism or other neurodivergence from one's experience of gender.

But *this* is what makes logic a space, a means, and a tool for freeing people rather than oppressing them, by giving them a space to exploit where their natural tendencies and inclinations lie, independent from their gender—it is precisely logic's attractiveness to people for reasons that not only have nothing to do with whether they are a man or a woman, but may in fact be rooted in the ways in which they are a man or a woman, or indeed neither.

Plumwood agrees with Nye that logic can be used as a tool of oppression when it involves reductive negation, partitioning things into X and non-X; it is oppressive because of the way it erases different ways of being a non-X. But just as classical negation erases the many ways in which there can be more than one way to be a non-X, something that neither Plumwood nor Nye recognize is that in many cases, there is more than one way *to be an X*.

It is less the dualistic nature of negation and the way it homogenizes the "non-Xs" that makes classical logic a tool of oppression and more the process of homogenization itself—suppose instead of dividing the world into "men" and "non-men," we divide it into "women" and "non-women". Just as there is harm in the homogenizing nature of the negation "non-men," there is harm in the homogenizing nature of "women". Just as there are many ways in which something can be not X, there are also many ways of being an X: Just as there are many ways in which one can be a non-woman, there are also many ways of being a woman! The trap that many people, philosophers included, fall into is insisting that there is one way of being a woman (or being a "woman in philosophy") and excluding those who do not perform in or inhabit that way.

Logic doesn't enforce any of these binaries or dichotomies or exclusions upon us. Logic has the potential to give the space to be maximally inclusionary. It is precisely the extent to which logic is open and available to women, including (especially!) those who don't necessarily perform femininity the way that (some) feminists may want us to, that logic can provide a space of liberation.

Let's go back to our original questions, about who is logic for, who gets to be a logician or count as a logician, who is excluded from logic, and by what means. Asking these questions matters because it forces us to either decide to exclude people, or decide no one should be excluded. If we decide no one should be excluded, we have to look at our exclusionary practices:

Where do they come from? What are the assumptions they are rooted in? How do we combat them? These are questions I cannot answer here (though which I intend to address in future work), but the first step in answering them is recognising that there is nothing *inherently* exclusionary in the study and practice of logic, and that while we must recognize that it can be used as a tool of oppression, this is not essential to logic and that it can also be used as a tool of liberation.

References

Cooper, K., Smith, L. G. E., & Russell, A. J. (2018). Gender identity in autism: Sex differences in social affiliation with gender groups. *Journal of Autism and Developmental Disorders, 48*, 3995–4006.

Duvekot, J., van der Ende, J., Verhulst, F. C., Slappendel, G., van Daalen, E., Maras, A., & Greaves-Lord, K. (2017). Factors influencing the probability of a diagnosis of autism spectrum disorder in girls versus boys. *Autism, 21*(6), 646–658.

Eckert, M., & Donahue, C. (2020). Towards a feminist logic: Val Plumwood's legacy and beyond. In D. Hyde (Ed.), *Noneist Explorations II: The Sylvan Jungle* (Vol. 3, pp. 424–448). Dordrecht: Synthese Library.

Ganeri, J. (2001). Introduction: Indian logic and the colonization of reason. In J. Ganeri (Ed.), *Indian Logic: A Reader* (pp. 1–25). London: Curzon.

Gould, J. (2017). Towards understanding the under-recognition of girls and women on the autism spectrum. *Autism, 21*(6), 703–705.

Hale, E. E. (1913). Women and logic. *The North American Review, 198*(693), 206–217.

Lai, M.-C., Lombardo, M. V., Auyeung, B., Chakrabarti, B., & Baron-Cohen, S. (2015). Sex/gender differences and autism: Setting the scene for future research. *Journal of the American Academy of Child & Adolescent Psychiatry, 54*(1), 11–24.

Nye, A. (1990). *Words of Power: A Feminist Reading of the History of Logic.* New York: Routledge.

Plumwood, V. (1993). The politics of reason: Towards a feminist logic. *Australasian Journal of Philosophy, 71*(4), 436–462.

Sara L. Uckelman

van Schalkwyka, G. I., Klingensmith, K., & Volkmar, F. R. (2015). Gender identity and autism spectrum disorders. *Yale Journal of Biology and Medicine*, *88*, 81–83.

Sara L. Uckelman
Durham University, Department of Philosophy
England
E-mail: s.l.uckelman@durham.ac.uk

www.ingramcontent.com/pod-product-compliance
Lightning Source LLC
Chambersburg PA
CBHW070919180426
43192CB00038B/1833